24 0438878 6

Imperial College
of Science, Technology and Medicine

Physics Department Library

Class Number..........

NEW MODES OF PARTICLE ACCELERATION—TECHNIQUES AND SOURCES

NEW MODES OF PARTICLE ACCELERATION— TECHNIQUES AND SOURCES

Santa Barbara, California   August 1996

EDITOR
Zohreh Parsa
Brookhaven National Laboratory

AIP CONFERENCE
PROCEEDINGS 396

American Institute of Physics   Woodbury, New York

Authorization to photocopy items for internal or personal use, beyond the free copying permitted under the 1978 U.S. Copyright Law (see statement below), is granted by the American Institute of Physics for users registered with the Copyright Clearance Center (CCC) Transactional Reporting Service, provided that the base fee of $10.00 per copy is paid directly to CCC, 222 Rosewood Drive, Danvers, MA 01923. For those organizations that have been granted a photocopy license by CCC, a separate system of payment has been arranged. The fee code for users of the Transactional Reporting Service is: 1-56396-728-6/ 97 /$10.00.

© 1997 American Institute of Physics

Individual readers of this volume and nonprofit libraries, acting for them, are permitted to make fair use of the material in it, such as copying an article for use in teaching or research. Permission is granted to quote from this volume in scientific work with the customary acknowledgment of the source. To reprint a figure, table, or other excerpt requires the consent of one of the original authors and notification to AIP. Republication or systematic or multiple reproduction of any material in this volume is permitted only under license from AIP. Address inquiries to Office of Rights and Permissions, 500 Sunnyside Boulevard, Woodbury, NY 11797-2999; phone: 516-576-2268; fax: 516-576-2499; e-mail: rights@aip.org.

L.C. Catalog Card No. 97-72977
ISBN 1-56396-728-6
ISSN 0094-243X
DOE CONF- 960836

Printed in the United States of America

Contents

Preface vii

Status and Future Directions for Advanced Accelerator Research 1
    R. H. Siemann

RF Sources for Future Colliders 11
    R. M. Phillips

Laser Acceleration in Vacuum 21
    J. L. Hsu, T. Katsouleas, W. B. Mori, C. B. Schroeder, and J. S. Wurtele

New Advances in Inverse Cerenkov Acceleration 31
    W. D. Kimura

Beam Driven Plasma Wake-Field Acceleration 41
    A. N. Skrinsky

The Open Waveguide Structure 53
    R. H. Pantell

Self-Modulation of High-Intensity Laser Pulses in Underdense Plasmas and Plasma Channels 61
    N. E. Andreev, L. M. Gorbunov, V. I. Kirsanov, and A. S. Sakharov

Self-Focused Particle Beam Drivers for Plasma Wakefield Accelerators 75
    B. N. Breizman, P. Z. Chebotaev, A. M. Kudryavtsev, K. V. Lotov, and A. N. Skrinsky

Laser Acceleration of Electrons: Zero to c in Less than Ten Microns 89
    D. Umstadter

Acceleration and Collision of Ultra-High Energy Particles Using Crystal Channels 95
    P. Chen and R. J. Noble

A Survey of Microwave Inverse FEL and Inverse Cerenkov Accelerators 105
    T. C. Marshall and T. B. Zhang

Requirements to Beam Emittances at Photon Colliders: Laser Cooling of Electron Beams 121
    V. Telnov

Parametric X-ray Radiation as Source of Pulsed, Polarized, Monochromatic, Tunable X-ray Beam 135
    Z. Parsa and A. V. Shchagin

Advanced Linacs for a Linac Ring Collider 145
    D. B. Cline

High Current Short Pulse Ion Sources 155
    K.-N. Leung

Positron Production in Multiphoton Light-by-Light Scattering 165
    C. Bula

Inverse Free Electron Laser Acceleration with a Square Wave Wiggler 179
    Z. Parsa and M. P. Pato

Schedule 190

List of Participants 195

Author Index 201

PREFACE

This Preface includes an introduction and a summary of a Symposium on "New Modes of Particle Acceleration: Techniques and Sources," which was held August 19–23, 1996 at the Institute for Theoretical Physics (ITP) in Santa Barbara. This was the first of the three symposia[1] hosted by the ITP and supported by its sponsor the National Science Foundation, as part of the 1st U.S. long term accelerator research program on "New Ideas for Particle Accelerators." The long term program and symposia were organized and chaired by Dr. Zohreh Parsa of Brookhaven National Laboratory/ITP. This Symposium provided a perspective on the future direction of the Advanced Accelerator Research.

The experimental study of elementary particles has become concentrated at a few large laboratories throughout the world because of the size and cost of the accelerator facilities needed for this work. For example, the Large Hadron Collider (LHC) at CERN, currently under construction, is 27 km in circumference and is being financed by the European membership of CERN, plus by contributions from non-member nations. An evolutionary approach to construction of ever higher energy colliders will only continue this trend towards high cost and large size.

Experimental particle physics cannot continue indefinitely along this path if it is to have a healthy long term future. Methods must be sought to reduce the size and cost of accelerators and thereby return scope and diversity to particle physics. Revolutionary changes in acceleration are needed. These could be new methods for colliding hadrons, new collider concepts such as muon colliders, or the use of new technologies, for example lasers or high frequency RF, in electron-positron linear colliders. Also new methods of experimentation such as compensated e^+e^- collisions (e.g., introduction of plasma at intersection point and four beam,...), new final focus concepts, or $\gamma-\gamma$ and $\mu^+\mu^-$ colliders should be investigated. These ideas and technologies should be considered to the degree in which they will bring revolutionary, not evolutionary, changes to accelerators and to the way collisions at ultra high energies can be achieved.

The ITP conference on New Modes of Particle Acceleration featured several presentations reviewing current progress in developing revolutionary accelerators based on laser driven plasma waves. In 1979, Dawson *et al.* proposed three basic laser plasma acceleration concepts; however only with the recent development of compact terawatt laser systems could these concepts be fully investigated in the laboratory.

The three proposed schemes were laser wakefield acceleration (LWFA), the plasma beat-wave accelerator (PBWA), and the self-modulated laser-wakefield accelerator (SMLWFA). In the LWFA, a single short laser pulse of length L excites a plasma wave of wavelength λ_p. In this scheme $L \simeq \lambda_p$. This method requires short, ≤ 1 pico-second, laser pulses of ultra high intensity $\gtrsim 10^{18}$W/cm^2 and could not be tested until chirped-pulse amplification (CPA) was used to create Table-Top Terawatt (T^3) lasers. Two papers on progress in T^3 technology based on CPA in solid state lasers were presented at the ITP Symposium by University of Michigan (UMI) and University of California at San Diego (UCSD).

The PBWA was proposed earlier as an alternative to LWFA because short-pulse, high-power lasers were not available. This approach employs two long pulse laser beams of slightly different frequencies, w_1 and w_2, such that $w_1-w_2 \simeq w_p$ the frequency of the plasma wave which is to be resonantly excited. PBWA experiments have been performed in Japan (ILE), the USA (UCLA), Canada (CRL), and France (LULI). The UCLA experiment observed the highest electron energy

---

[1] In addition to this symposium, a week long symposium was held on "Future High Energy Colliders" October 21–25, 1996. Some of the highlights of that meeting included discussions on the future direction of high energy physics by bringing together leaders from the theoretical, experimental and accelerator physics communities. A third symposium on "Beam Stability and Nonlinear Dynamics" was held on December 3–5, 1996 and dealt with some of the fundamental theoretical problems associated with accelerator physics.

gain, ~28 MeV [Clayton (UCLA) et al.], with an effective accelerating gradient of 2.8 GV/m. They plan to continue with PBWA experiments.

The most impressive advanced report at the Conference came in the area of self-modulated laser wakefield acceleration (SMLWFA). In this method, a laser pulse of length $L > \lambda_p$ is subdivided into a series of shorter pulses of length $\sim \lambda_p/2$ by its interaction with the plasma wave (which it created). This interaction creates a large amplitude (resonantly driven) plasma wave. This process requires a laser power greater than the critical level required for relativistic guiding of the laser field. The phase velocity of the guiding plasma wake can become relativistic for high enough plasma electron densities, $n_p \sim 10^{19}$ cm^{-3}, for example.

Experiments on SMLWFA have been performed in Japan (KEK), the US (LLNL, CUOS, NRL), and the UK (RAL). The latter experiment achieved impressive results: electron energy gains of $\gtrsim 44$ MeV and accelerating gradients $\gtrsim 100$ GV/m. Conventional accelerators are capable of accelerating gradients of ~100 MV/m. This experiment employed a 2.5 TW, 0.5 picosecond laser, producing an intensity of 10^{19} W/cm^2 and a plasma electron density of 10^{19} cm^{-3}.

The accelerated electrons in this experiment cover a wide range of energies from a few MeV up to the maximum. The theoretical limit for this experiment was ~70 MeV. The spectrometer was capable of measuring only up to 44 MeV. The normalized transverse emittance of any particular energy group was about 5π mm-mrad, which is on the order of the emittance of photo injector based linacs. However, the measured beam current was 10–100 times lower than that achieved with photoinjectors.

Although the reported accelerating gradients for SMLWFA are spectacular, they are achieved over short distances of the order of 100's of microns to millimeters. The size of the acceleration distance is determined by the diffraction limited Rayleigh length (of the region of minimum focal spot size). Various schemes for getting accelerating lengths greater than a few Rayleigh lengths were discussed at the Conference.

For example, optical guiding can be achieved with preformed lower density plasma channels produced by hydrodynamic expansion of ionized gas generated by another laser focused along the acceleration axis. Other approaches suggested were using laser blow out to create a low density hollow plasma channel or using acoustic wave channel formation. Relativistic focusing can provide optical guiding in the case of SMLWFA for laser power levels $P > P_c = 17(w/w_p)^2$ GW. Other limiting factors on accelerating lengths are electron detuning, (i.e., the length over which the accelerated relativistic electron outruns the plasma wave), and pump depletion, (i.e., is roughly the length over which the laser pulse gives up all its energy to the plasma wake).

All these SMLWFA experiments have accelerated background plasma electrons. Producing and injecting 20–100 femto-second electron bunches is a difficult challenge. Umstadter et al. propose to solve this problem by using two orthogonally propagating laser pulses, one along the acceleration direction as a plasma wave pump pulse and one as an injection pulse at right angles. This scheme is called LILAC, laser injection and laser acceleration. The University of Michigan is building an all optical accelerator to produce femto-second electron pulses with GeV energies.

Another approach discussed at the conference was that of Plasma wakefield accelerators, which are similar to LWFA, except that one or more relativistic electron beams are used to excite the accelerating plasma wave. The electron beam pulse must be shorter than the plasma wavelength in analogy with the situation for the LWFA. This concept was originally proposed by Fainberg in 1956. Enhancing the wakefield by using multiple electron drive bunches spaced at the plasma period was proposed in the original work on PWFA. The first PWFA experiment was performed by Berezin and co-workers in the early 1970's in the Ukraine. More recently, experiments were performed in the US (e.g., at Argonne Nat. Lab) and in Japan (at KEK).

At this conference Skrinsky and co-workers proposed a system design for a 1 TeV PWFA using a pre-ionized hydrogen plasma, which is driven by trains of electron bundles, which are made of 10 micro-bunches of 0.2 mm length. This system would employ a 10 GeV drive beam at 10 kHz rep rate with 2×10^9 electrons/bunch. Challenges involve the energy requirements of maintaining the hydrogen plasma channel. A PWFA test experiment is being proposed at INP, Novosibirsk with a goal to reach more than 0.5 GeV/m over several tens of cm.

The exciting experimental results on high gradient laser plasma acceleration and the successful modeling and simulation of these systems gives great confidence that these concepts are beginning to be understood well enough to plan the next stage of accelerator development. Such second generation plasma accelerators would address issues of beam quality. It was proposed that a modest near term goal be the production of 100 MeV electrons with an energy spread of ~5%, normalized emittance of $\lesssim 10\pi$ mm-mrad and number or particles per bunch $\gtrsim 10^8$ (T. Katsouleas, University of Southern California (USC)). This would require injecting pre-bunched beams with $\tau_{bunch} < 60$ fsec. Another scheme (D. Umstadter, University of Michigan (UMI)) plans for all optical laser accelerators to produce femto-second electron pulses at the GeV level.

The field of laser plasma acceleration should continue to benefit from research on high power, short pulse lasers. The potential for applications (in other fields) of these high power lasers helps to maintain interest and support for this area of research. Some applications of these lasers include ultrafast x-ray sources for time-resolved diffraction studies of phase transitions in materials, time-resolved absorption spectroscopy, and high resolution, time-gated radiology. For example, the latter application could, in principle, result in improved resolution and lower patient dose for mammography. These lasers are also of interest in studies of ultra-dense plasma physics and highly relativistic laser-matter interactions.

Among the other subjects treated were power sources, such as RF Sources (R. Phillips, Stanford Linear Accelerator Center (SLAC)), Laser as a power source (G. Mourou, UMI), Pulsed Power Sources (M. Gunderson, USC); Advanced Accelerator Schemes such as Laser Acceleration (W. Mori, University of California, Los Angeles (UCLA)), Two Beam Accelerator (S. Yu, Lawrence Berkeley Lab (LBL)), Inverse Free Electron Laser & Free Electron Lasers (C. Pellegrini, UCLA), Inverse Cerenkov (W. Kimura, STI Optronics), Open Waveguide Structure for Laser Acceleration (R. Pantell, SLAC); Beam Cooling (A. Skrinsky, Institute of Nuclear Physics, Novosibirsk (BINP)); Laser Cooling (A. Sessler, LBL); Crystal Accelerator (P. Chen, SLAC).

There were other presentations including parametric x-radiation (A. Shchagin, Kharkov Inst. of Physics and Technology, Ukraine), Femto-second x-ray pulses (A. Zholents, LBL), Advanced Linear Accelerator development (D. Cline UCLA). Beam Sources such as High current short pulse Ion Sources (K. Leung (LBL)), High Intensity Neutron Spallation Sources (R. Macek, Los Alamos National Lab (LANL)), High Intensity Muon Source; Beam Dynamics and emittance, etc.

Laser Plasma issues (Esarey, Naval Research Lab (NRL)), Self Modulation of Intense Laser Pulses in plasma Channels (N. Andreev, Russian Academy of Science (RAS)), Production of Ultra Short Laser Pulses (C. Barty, UCSD), Short Bunch Injection, Synchronization and Acceration in Laser Wakefields (D. Umstadter, UMI), update on Laser Plasma experiments (C. Joshi, UCLA) also were discussed.

The symposium included two unique discussion sessions on "Laser Plasma Based Acceleration" and on "New Advances and Basic Issues" in which participants presented and clarified their views on outstanding problems and topics presented at this conference. This forum provided new and valuable input for future direction and developments in this field. There was a great interest and request by participants to write up a "white paper" on the future direction of the advanced accelerator research. We thank those who have sent in contributions and suggestions on what should be included in the write up of the "white paper."

In terms of a white paper, Umstadter suggests that the critical issues to be studied are (1) what are the means of injecting electrons with femtosecond precision, (2) what are the means of creating plasma channels many Rayleigh lengths long, and (3) can the electron beam properties preserved through multiple synchronized accelerating stages? Of course, it must also be demonstrated that electron beams can be accelerated in a single stage with suitable properties for high energy physics.

E. Esarey noted that overall, the new laser technology has led to a rapid outgrowth of new experimental results on the LWFA. These results were obtained relatively easily (in a few months) and use the simplest possible configurations. Both the laser technology and the LWFA experimental results are improving rapidly—nearly on a monthly basis. The major experimental research on the LWFA in the US is at NRL and Michigan (the plasma beat wave accelerator is at UCLA). Experimental programs are also being developed at Texas, Maryland, LBL, Livermore and UCSD. Funding in the US is modest at best, funded partially by DOE and internal lab funds. (Livermore announced a new lab initiative this summer.) Japan has announced this past fall (JAERI Japanese Atomic Energy Research Institute) the start of a large program ($100 million/year excluding salaries, and plan on hiring 200 full time scientists over the next five years) on the development of short pulse lasers and their applications, including the LWFA. England (Rutherford) and France (LULI) also have significant programs. If the US is to stay competitive in this field, a higher commitment of research funds is necessary.

There were many other very interesting comments from the participants, but due to time and space limitations they are not included here.

The Symposium started with a defining perspective presentation by R. Siemann (SLAC) and ended with a summary and closing presentation by Z. Parsa (BNL). The symposium was a success, with a very interesting program and an overwhelming active group of expert participants.

I would like to thank all the authors for providing the write-ups of their talks. For one reason or another several speakers were not able to provide the write-ups of their talks. I nevertheless thank them for their stimulating talks and participation in the symposium. In most cases, copies of the transparencies from their talks can be found in the report BNL-52523. For a complete list of presentations, see the program schedule given in the Appendix. I would like to thank the advisory committee, all speakers, conveners, and participants for making the symposium a unique and stimulating experience. I would also like to thank Clifford and other participants for providing the photographs and special thanks to the ITP Director, Manager, and staff for providing a beautiful setting and making sure the meeting ran smoothly.

Zohreh Parsa
Chairperson, Symposium
Institute for Theoretical Physics,
UCSB, Santa Barbara, California
Brookhaven National Laboratory, Upton, New York

Status and Future Directions for Advanced Accelerator Research†

R. H. Siemann*

Stanford Linear Accelerator Center,
Stanford University, Stanford, CA 94309

ABSTRACT: The relationship between advanced accelerator research and future directions for particle physics is discussed. Comments are made about accelerator research trends in hadron colliders, muon colliders, and e+e- linear colliders.

COLLIDERS AND HIGH ENERGY PHYSICS

The mass scale of interest to particle physics is the range of ~ 0.5 to 2 TeV where electroweak symmetry is broken. Experiments at colliders with high enough energy are expected to detect evidence of electroweak symmetry breaking and to shed light on the symmetry breaking mechanism. Is it the classic Higgs phenomena, supersymmetry, strong coupling, or something else? History suggests that discovering the origin of electroweak symmetry breaking will also raise questions about subjects unknown today.

The Large Hadron Collider (LHC) is a technically proven and funded project that could reach high enough energy and luminosity for the study of electroweak symmetry breaking, and the NLC, JLC and TESLA, linear colliders being designed for center-of-mass energies E_{CM} = 0.5 to 1.5 TeV, promise an unrivaled environment for the study of this phenomenon. The sizes and costs of these colliders raise questions that are at the heart of the future of particle physics

1. Are the colliders and detectors needed for the study of electroweak symmetry breaking affordable? The costs of these facilities are modest on the scale of many governmental activities, so the issue is whether our elected representatives decide that high energy physics pursued at this scale is or is not in the national interest. The SSC was started when they decided it was, but that project was terminated when their opinion changed. CERN and the LHC may be facing problems of the same nature with the discussion of budget cuts initiated by the German government.

International collaboration on the design, construction and operation of large colliders is the proposed solution to the high cost of these facilities. The cost per country is reduced, but the involvement and commitment of each country is reduced also. Will one or two large colliders located somewhere in the world meet the needs of the governments that support particle physics, the institutions

---

* Work supported by the Department of Energy, contract DE-AC03-76SF00515.

that commit faculty and staff to this scholarly field, and the physicists who perform the research? If these needs are not met, support for and interest in the field could drop precipitously. The discussion of international collaboration has concentrated on the cost reductions without much consideration of these needs and the consequences of not meeting them.

2. Do we have the technology and accelerator physics to move on to the next energy scale? The colliders of today are based on a combination of principles, technologies, and accomplishments that has led to many of the past discoveries in particle physics and has placed the field on the verge of studying the Higgs phenomenon. However, these accomplishments are not enough for the future. We are at the limit of affordability, and an extension of present techniques is not a way to reach the next energy scale.

High energy physics based on an extrapolation of present trends will be a field posing exciting scientific questions but with few opportunities to explore them and with high costs. Reduced opportunities and remoteness from universities, laboratories and nations could reduce institutional and national commitment to particle physics, and the LHC and the next generation of linear collider could be the last major facilities constructed for this science.

This demise of particle physics seems inevitable unless there is a revolutionary change in particle accelerators that reduces costs. This must be a revolution comparable to that which replaced vacuum tubes with integrated circuits and telephone wires with fiber optics and cellular facilities. These are examples of inventions that were so dramatic that new, previously undreamed of ideas became possible. Particle physics must have inventions of comparable impact.

COLLIDERS

Characteristics

Colliders have characteristics that describe the particle physics potential: luminosity, center-of-mass energy, lepton or hadron beams, backgrounds, interactions per crossing, energy spread, collision spot size, etc. Some of these such as luminosity and center-of-mass energy are the raison d'être, and these should be the goal of accelerator development.

Others have major impact on experiments, and accelerator physicists try to make those impacts as favorable as possible. Backgrounds and interactions per crossing are examples. These can be given less emphasis, perhaps even ignored, when revolutionary changes in accelerators are required. Comparable changes in experimentation are going to be necessary also. The choice is as stark as it is for future colliders - work on innovations in experimentation or the survival of particle physics is in question.

Topology

The general topology of a collider has a particle source, accelerator, storage system, and collision system. The two examples given in Table 1 show

Table 1. General Collider Topology and Two Examples

General Topology	SLC	Tevatron
Particle Source	Polarized Gun, Damping Rings	Ion Source, $\bar{p}$ Cooler and Accumulator
Accelerator	SLAC S-Band Linac	Booster, Main Ring, Tevatron
Storage System	------	Tevatron
Collision System	Final Focus	High β Quadrupoles and Interaction Region

that *i)* the SLC has three of four of these systems and *ii)* the functions are combined or closely connected in the Tevatron.

A collider must have most of these systems, and they must work together, complement each other, and the properties of one system can strongly influence other systems. Two examples of that the dominant role of $\bar{p}$ production and cooling in all of the other Tevatron systems, and need for flat beams at the collision point of a linear collider determining many of the parameters of the damping rings and accelerator. While much of this is obvious, it is often ignored in the advanced accelerator community which can become fascinated with an aspect of performance without considering possible functioning as a collider.

OLD AND NEW INVENTIONS

The accelerators and colliders of today are based on:

1) ***Great principles*** of accelerator physics: phase stability, strong focusing, and colliding beam storage rings;
2) ***Dominant technologies***: superconducting magnets, high power RF production, and normal and superconducting RF acceleration;
3) Many other ***substantial accomplishments*** in accelerator physics and technology: non-linear dynamics, collective effects, beam diagnostics, etc.;
4) Years of ***experience*** with operating colliders. This is closely related to the previous element. Overcoming performance limits has often required development of sophisticated theories, experiments, or instrumentation.

A change in the future of high energy physics will require inventions and new ideas of comparable importance to the great principles and dominant technologies. These must encompass both accelerator physics and technology to have the needed impact.

Particle physics is only a small part of science, and these critical ideas may arise in other contexts and have other driving forces including market forces. The accelerator community needs to be aware of developments throughout science and technology and constantly be considering the application of new developments to particle physics. High peak power lasers are a clear example. These devices are being developed for a wide range of scientific and commercial applications, and in the process devices with enormous potential for producing high acceleration gradients are becoming available.

HADRON COLLIDERS

This is the first of three sections that deal with the colliders that could have a role in the future and with issues related to them.

High energy hadron colliders are a proven way to reach the energy scales of interest to high energy physics. Unfortunately the costs of today's technology are prohibitive for thinking about future extrapolations, and the focus of hadron collider development has to be cost reduction.[1] The SSC can be used to understand costs and to identify areas with potentially significant savings. The Appendix shows that the superconducting magnets of the collider ring were almost half of the SSC cost. This is the area where there must be significant savings.

There is extensive experience at the Tevatron, HERA, RHIC, SSC, and LHC with 4 - 8 T $\cos\theta$ magnets. This is the technology determining the present energy frontier. However, since this type of magnet is well developed, it is unlikely to be the basis for the qualitative changes needed in the future. Directions that hold promise for such changes are low-field, superferric magnets and high temperature superconductors.

The low-field superferric magnet[3] addresses many of the costly aspects of higher field magnets. The geometry is simple with a single conductor placed in a low magnetic field region. The principle disadvantage is that the field is low, $B \le 2T$, because iron is used to shape it. As a result the collider must be large, several hundred km in circumference, and that has consequences for beam stability, stored beam energy, etc.[2] Magnet development together with further work on the consequences of low field should indicate whether this is a viable and cost effective idea.

Table 2 is a comparison of superconductors which shows the high critical magnetic fields and critical temperatures of the high T_c superconductors BSCCO and YBCO. These intrinsic properties make the materials attractive, but the superconductor volume fraction, the mechanical properties, and the production of material must be improved. There will be help from outside high energy physics because of potential commercial applications. In addition to improving the

Table 2. Comparison of Superconductors (Ref. 2)

Property	NbTi	Nb$_3$Sn	BSCCO-2223	YBCO
Upper Critical Magnetic Field (T)	15	25	~ 100	~ 100
Critical Temperature (K)	9.5	18	110	92
Critical Current Density* (kA/mm^2)	2 - 2.3	1 - 2.4	< 0.9	< 2.4
Superconductor Volume Fraction (%)	40 - 50	35 - 40	35 - 40	~4
Conductor Type	multifilament wire	multifilament wire	multifilament tape	microbridge
Mechanical Property	Ductile	Brittle	Brittle	Brittle
Longest Piece Made	~ 10 km	> 1 km	~ 1 km	~ 10 mm

* The magnetic fields and temperatures for the critical current densities are: NbTi - 7 T & 4.2 K or 10 T & 1.8 K; Nb$_3$Sn - 10 T & 4.2 K; BSCCO - 20 T & 20 K; YBCO - 20 T & 77K.

materials, there need to be ideas about how high T_c superconductors might be used in an accelerator magnet. High field magnets are not attractive at the present time, and the superferric magnet appears to be the only possibility.

MUON COLLIDERS

There are two premises leading to the interest in muon colliders for ultra-high energies. The first is that lepton-lepton collisions are necessary because the radiation damage to detectors at hadron colliders will be unacceptable, and the second is that beam-beam effects are a critical flaw of linear e^+e^- colliders. These are strong criticisms of hadron and linear e^+e^- colliders, and they deserve being addressed. Possible answers could include *i)* novel experimental techniques, *ii)* changes to the linear collider paradigm, and *iii)* the muon collider.

The muon collider consists of a high intensity proton synchrotron, a muon production system, ionization cooling stages, accelerators capable of bringing the beams to collision energy rapidly, and a collider ring.[4] A system approach has been taken to the design of a muon collider with all of the elements of the general topology of Table 1 being considered at the same time. Since each of the major component systems has significant technological and/or beam dynamics issues, this approach optimizes the collider concept and focuses research on critical issues.

Some people believe that since a complete collider concept is being discussed, the muon collider has moved from the realm of advanced accelerator research to that of project oriented research. This is not the case. The muon collider poses research questions in many fundamental areas of accelerator physics and technology. Beam current limits in proton synchrotrons and ionization cooling are two examples. The muon collider provides a context for the study of this accelerator physics just as an e^+e^- linear collider and a hadron collider provide ones for research in high gradient acceleration and high T_c superconductors, respectively.

ELECTRON-POSITRON LINEAR COLLIDERS

There is no complete concept for a 5 - 10 TeV e^+e^- linear collider, but there are several issues of clear importance.

Limitations of the Beam-Beam Interaction

The expressions for luminosity, $\mathcal{L}$, beam power, P_B, and the number of beamstrahlung photons per incident particle, n_γ, can be combined to give

$$\mathcal{L} \approx \frac{1}{8\pi\alpha r_e} \frac{P_B n_\gamma}{E \sigma_y} . \tag{1}$$

The beam energy is denoted by E, and the vertical beam size, σ_y, is assumed much smaller than the horizontal beam size, σ_x. The other quantities in the equation are: α = fine structure constant; and r_e = electron classical radius. This equation shows the well-known trade-offs between beam power, vertical spot size and beamstrahlung. The factor n_γ in the numerator is taken as a measure of backgrounds produced by the beam-beam interaction. Increasing the collision point electromagnetic fields increases beamstrahlung and luminosity. If there is a limit on beamstrahlung from detector backgrounds, there is a limit on luminosity.

This expression is valid when the collision point electromagnetic fields are much less than the critical magnetic field, $B_C = 4.4 \times 10^{13}$ G. When the fields are comparable to B_C, phenomena such as coherent pair production increase backgrounds dramatically.[5] The parameter Y,

$$Y = \frac{\gamma B}{B_C} \approx \frac{r_e^2 \gamma N}{\alpha \sigma_z (\sigma_x + \sigma_y)} \qquad (2)$$

($\gamma = E/mc^2$; B = collision point magnetic field; N = number of particles per bunch; σ_z = bunch length), is usually kept Y < 0.3 in linear collider designs. This becomes increasingly difficult at high energies because of *i)* the direct proportionality to γ, *ii)* high gradient structures have short wavelengths and the bunch length must be a small fraction of the wavelength, and *iii)* the need for small σ_y together with limits on σ_x/σ_y from beam optics.[6] If Y < 0.3 is necessary, this could be the critical flaw of e^+e^- linear colliders mentioned earlier in the muon collider section. However, there are several possible ways to deal with the limitations of the beam-beam interaction within the linear collider concept.

The first is to *ignore it*. This may be wishful thinking, but perhaps it isn't. High field Quantum Electrodynamics with Y ~ 1 has been studied experimentally in laser - electron beam interactions,[8] but there is no experience with beam-beam related backgrounds in a linear collider. Real life will be different than the Monte Carlos studied to date which have considered backgrounds in an extrapolation of today's high energy collider detectors. A compelling multi-TeV linear collider concept will spark creativity in the experimental physics community, and innovative approaches to experimentation could emerge.

The second approach to the limitations of the beam-beam interaction are to *avoid them* with a different collision paradigm. One possibility is photon-photon rather than e^+e^- collisions.[9] There are no issues of beamstrahlung or coherent pair production in a photon-photon collider, and the dominant problem is the configuration near the collision point. Accelerated electrons have to be converted to photons by Compton scattering with an intense laser, and this conversion point must be close to the collision point for high luminosity.

The other possibility of a different collision paradigm is plasma[10] or beam compensation where fields at the collision point are reduced by neutralization. There would be substantial backgrounds from interactions in a plasma if one were used to neutralize the collision. The creativity of experimentalists would be required to deal with them. Compensation with beams would require overlapping electron and positron beams. Efficient generation and control of such beams together with the stability of the compensated configuration are all problems to be solved. There are ideas for this.[11]

Harnessing the Potential of the Laser

High peak power lasers are a breakthrough technology, and exploiting their enormous potential for particle acceleration is one of the major challenges for accelerator physics research. They have found use already for the generation of low emittance beams in laser driven RF guns, and they could have a role in generation of power at high frequencies.[12] However, the primary interest has to be with the high gradients possible in a laser driven accelerator.

Different laser driven accelerators have been studied both theoretically and experimentally. Far field accelerators (of which the Inverse Free Electron Laser (IFEL) is the most prominent) couple to the transverse electric field of the laser by giving particles a transverse component of motion. This motion generates synchrotron radiation which limits the beam energy. Far field accelerators could find application as injectors or bunchers, but the energy limit makes them relatively uninteresting for high energy physics.

There have been many ideas for direct acceleration of a beam with a laser by using structures to give a longitudinal component to the laser field. Structures with features comparable to the laser wavelength are similar to RF driven linacs. Lithographic techniques could be used for fabrication, but there will be stringent limitations on accelerated charge from wakefields. These limitations are so severe that interest in this type of structure has dropped substantially. Current interest is focused on structures with the features in at least one dimension large compared to the laser wavelength. Crossed laser beams[13] and a structure similar to the open optical waveguide are being considered.[14] Both promise gradients ~ 1 GeV/m with substantially lower wakefields than optical renditions of RF linacs.

The highest acceleration gradients achieved to date have been with laser driven plasma accelerators. Plasma waves can be excited resonantly in the laser beatwave accelerator or by the excitation of a wakefield with a short, high intensity laser. The laser pulse is self-modulated when the pulse is long compared to the plasma wavelength. Gradients of ~ 100 GeV/m have been observed in the latter configuration.[15] This type of result has attracted widespread interest, and the field of laser driven plasma accelerators is moving on to achieving this acceleration over long distances, staging of multiple accelerators, and beam quality and stability. When these have been successfully addressed the plasma accelerator will attract the interest of the mainstream accelerator community.

Short Wavelength & High Gradient Limits of Metallic Structures

The SLC has an RF wavelength of 10.5 cm and an accelerating gradient of $G \sim 20$ MeV/m. While there is a variety of RF technologies being considered for a next generation of linear collider, the tendency is towards shorter wavelengths and higher gradients. A 5 - 10 TeV collider could be possible by going even further in this direction to mm wavelengths and GeV/m gradients.

The arguments for this include energy efficiency, which for a fixed gradient and number of particles is proportional to λ^{-2}, and the dependence of gradient on wavelength. The dominant phenomena limiting gradient at 1 - 10 cm wavelengths are *i)* capture and acceleration of dark current and *ii)* RF breakdown. Dark current capture depends on wavelength as $1/\lambda$.[16] Loew and Wang[17] have measured RF breakdown at a fixed pulse length of 1 μs and different frequencies. They find that the breakdown gradient is proportional to $\lambda^{-1/2}$. Correcting for reduced pulse length at shorter wavelengths, Wilson estimates that the gradient limit from RF breakdown is proportional to $\lambda^{7/8}$.[16] These are empirical results, and, while further research is needed to clarify underlying mechanisms, they argue for short wavelengths.

There are several disadvantages of short wavelengths. Longitudinal and transverse wakefields scale as $1/\lambda$ and $1/\lambda^3$, respectively. New ideas for aligning and stabilizing accelerating structures and beams are needed. Recent work on structure alignment based on detecting RF induced in deflecting modes may provide a basis.[18] There is a possible gradient limitation from pulsed heating. This is thought to scale as $1/\lambda^{1/8}$,[11] but the experimental information about pulsed heating in RF systems is contradictory. An experiment studying pulsed heating in RF systems is planned.[19] Structures and filling times get shorter with shorter wavelength, and the peak power per meter depends on gradient and wavelength as $G^2 \lambda^{1/3}$ [16] The consequences are that new RF power sources and pulse compression techniques are sure to be required. These problems must be solved for short wavelength, high gradient RF to be viable.

CONCLUDING REMARK: ACCELERATOR IR&D

The future of high energy physics and successful accelerator *Invention, Research and Development (IR&D)* are one and the same. The last three sections have discussed and commented on some of the current directions for advanced accelerator research in hadron, muon, and linear colliders for future generations of high energy physics colliders. Most of the ideas are not the revolutionary ones that are needed. However, my hope is that the combination of motivated, intelligent people and a supportive atmosphere will produce the critical insight that is so badly needed.

REFERENCES

† This paper has been submitted to the Proceedings of the 7th Workshop on Advanced Accelerator Concepts held in Lake Tahoe, CA
1. Much of the current thinking about high energy hadron colliders can be found in ref. 2.
2. G. Dugan, P. Limon and M. Syphers, "Really Large Hadron Collider Working Group Summary", Proc. of 1996 Snowmass Workshop..
3. G. W. Foster and E. Malamud, "Low-Cost Hadron Colliders at Fermilab", Fermilab TM-1976 (1996).
4. R. B. Palmer, A. Tollestrup and A. Sessler, "Status Report of a High Luminosity Muon Collider and Future Research and Development Plans", Proc. of 1996 Snowmass Workshop.
5. P. Chen, AIP Conf Proc **184**, 633(1989).
6. These general statements are supported with more detail in ref. 7.
7. S. Chattopadhyay, D. Whittum, and J. Wurtele, "Advanced Accelerator Technologies A Snowmass '96 Subgroup Summary", Proc. of 1996 Snowmass Workshop.
8. C. Bula *et al*, PRL **76**, 3116 (1996).
9. S. Chattopadhyay and A. Sessler editors, NIM **A355**, 1 (1995).
10. A. M. Sessler and D. Whittum, AIP Conf Proc **279**, 939 (1993).

11. D. Whittum presentation at 1996 Snowmass Workshop. See ref. 7.
12. W. Budiarto et al, "High Intensity THz Pulses at 1 kHz Repetition Rate", submitted to JQE (1996).
13. Y.-C. Huang presentation at 1996 Snowmass Workshop. See ref. 7.
14. R. Pantell, presentation at the 1996 Free Electron Laser Conference.
15. A. Modena et al, IEEE Trans Plasma Sci **24**, 289 (1996).
16. P. Wilson, SLAC-PUB-7256(1996).
17. G. A. Loew and J. W. Wang, SLAC-PUB-5320 (1990).
18. M. Seidel et al, to be submitted to PRL.
19. D. Pritzkau et al, private communication.
20. T. Elioff, "A Chronicle of Costs", SSCL-SR-1242 (April, 1994).
21. Table 6-2 of ref. 20.
22. Tables 6-3 and 6-4 of ref. 20.

APPENDIX: SSC COST ANALYSIS

While a detailed SSC cost analysis is complicated because project evolution and schedule changes had large impacts the cost,[20] the "Site Specific Conceptual Design" can be used to show relative costs. From Tables A-1 and A-2 one sees that 52% of the Total Project Cost (TPC) was in the accelerator system with the collider accounting for 42% of the TPC. When it is assumed that project management, contingency, R&D, and administrative and technical support should be apportioned according to system costs rather than appearing as separate items in the budget these percentages become 75% and 61%.

The accelerator systems (not including the magnets), superconducting magnets, and conventional systems of the collider are 17%, 44% and 10% of the TPC, respectively. Almost one-half of the cost is associated with the collider ring superconducting magnets.

Table A-1. SSC Site Specific Conceptual Design Costs* [21]

Category	SCDR Costs FY90$
Construction	
1.0 Technical Systems	2,986,400,000
2.0 Conventional Systems	1,051,500,000
3.0 Project Management	48,700,000
Contingency	753,000,000
Construction Subtotal (TEC)	*4,839,600,000*
Other Program Costs	
4.0 R&D, Pre-Operations, Administrative and Technical Support	975,900,000
5.0 Experimental Systems	752,100,000
Other Subtotal	*1,728,000,000*
Total Project Cost (TPC)	**6,567,600,000**

* These numbers correspond to a proposed actual year cost of $7,836,600,000 which was increased to $8,249,000,000 after reviews by the Department of Energy.

Table A-2. SSC Accelerator Technical and Conventional Systems[22] (1)

System	Accelerator Systems (2)	Conventional Systems	System Cost (3)		% of TPC (3)	
Linac	37	3	40	(58)	0.6	(0.9)
LEB	42	5	47	(68)	0.7	(1.0)
MEB	113	35	147	(212)	2.2	(3.2)
HEB	326	74	400	(576)	6.1	(8.8)
Injector	*518*	*117*	*635*	*(915)*	*9.7*	*(13.9)*
Collider	2,304	464	2,768	(3987)	42.1	(60.7)
Accelerators	*2,822*	*581*	*3,403*	*(4901)*	*51.8*	*(74.6)*

Notes: 1. Costs in FY90 M$. 2. Including superconducting magnets which are $1,668M$ of the collider cost. 3. The numbers in ()'s indicate costs and percentages with project management, contingency, R&D etc. allocated in proportion; (cost) = cost ¥ [1 + (48.7+753.0+975.9)/(2986.4 + 1051.5)].

RF SOURCES FOR FUTURE COLLIDERS

Robert M. Phillips

Stanford Linear Accelerator Center
Stanford University Stanford, California 94305

Abstract. As we push particle colliders to 1-TeV center-of-mass collision energy and beyond, we require much more from our RF energy sources, both in terms of the RF performance and the number required for a given machine. In order to conserve real estate, the operating frequency of future colliders is apt to be higher than the S-band used for the SLAC SLC. It is this inevitable trend toward higher frequencies which presents the source designer with the greatest challenge. This paper is about that challenge. For reasons which will become clear, as we go to frequencies substantiallly above X-band, we will require sources other than klystrons, probably of the type referred to as "fast-wave devices," such as FEL or gyro-based amplifiers, or two-beam accelerators. Because these are discussed elsewhere in this conference, I will stick to the klystron as my model in describing the challenges to be overcome, as well as the criteria which must be met by alternative sources for new accelerators.

THE CHALLENGE TO THE KLYSTRON

SLAC'S design for the NLC, operating at 1 TeV, requires approximately 10,000, 75-MW, X-band klystrons with a pulse of 1.2 µs and a rep rate of 120 Hz. First, let's consider what properties the source, whatever it is, is required to have. Keep in mind we are talking about 10,000 units. These include:

Klystron-like RF performance: For phase control purposes, an amplifier rather than an oscillator is required. High gain, preferably more than 50 dB is desirable, or the cost just gets pushed down to the driver level. About 1% bandwidth is required for pulse compression.

Reliability/availability: One wants every tube to turn on and operate whenever required. When a tube does misbehave, the accelerator has to be designed to go on without it.

Long life: Our S-band klystrons have a meantime-to-failure of 50,000 hours, which is considered quite extraordinary by microwave tube standards. It is impressive by almost any standard. For instance, an automobile with 50,000 hours life averaging 40 MPH would provide 2 million miles of useful transportation.

However, with 10,000 tubes we have to do about this well. Even at this level we will be replacing 1,200 tubes per year. That's 100 per month, or 3 1/2 change outs per day. Obviously, the change out procedure has to be straightforward and require little time. We might be able to live with double this rate of change outs, but 25,000 hours meantime-to-failure is pretty much the minimum acceptable.

High efficiency: The efficiency of the source directly impacts wall plug power. Our S-band klystrons operate at about 42% efficiency and this is not good enough for a 10,000 tube machine. We have set a goal of 60% efficiency for the X-band sources. This is further complicated by the requirement for a beam-focusing solenoid which, in the case of the X-band tube, consumes about half as much power as the klystron itself. We found a way of completely eliminating the beam-focusing power by using permanent magnets, which will be described later.

Low cost: If the source has high production cost designed into it, it simply won't fly. At SLAC we have begun a major effort to cost-reduce our X-band klystrons with a goal of $30,000 per klystron in production quantities. This effort includes the detailed design of the required factory.

MATCHING THE KLYSTRON TO THE ACCELERATOR

Figure 1 illustrates the RF pulse the physicist requires for the accelerator, 300-MW peak power with about 0.3 μs pulse length. Also shown is a pulse which is far more compatible with the output of a practical klystron, a 75-MW pulse with a maximum pulse length of 1.5 μs. This combination, in fact, is what we will use in the test accelerator for the NLC. The transformation is achieved using a pulse compressor which is basically a long, low-loss waveguide that is hooked up in a hybrid feed fashion so that it is first filled by the klystron and then made to empty when the phase of the klystron is reversed, giving a short, high-amplitude output pulse. The energy difference between the two pulses is lost because the filling and emptying process is not perfect.

I have been asked, "Why not avoid the pulse compression losses by simply designing a klystron which produces 300 MW of power?" This is probably doable, but in the real world it too would be inefficient and not very practical. The beam voltage would need to increase from about 500 kV to about 850 kV, which makes for a very difficult modulator design problem.

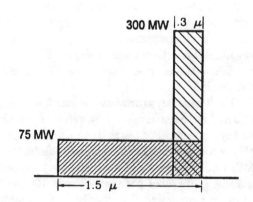

FIGURE 1. Matching Klystron Output to Accelerator Need

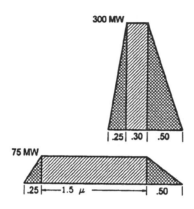

FIGURE 2.. Short Pulse Reduces Modulator Efficiency

Figure 2 shows why one does not really gain in efficiency. It is a cartoon of a typical modulator pulse with a 1.5 μs flattop which has a 0.25 μs rise time and a 0.5 μs fall time. At the higher voltage with the shorter flattop condition, it still has at least a 0.25 μs rise time and a 0.50 μs fall time, which now represents a significantly larger portion of the total energy, hence the inefficiency.

All would be well, if instead of a line-type modulator, one could use a DC voltage supply and a low-voltage, high-impedance control electrode like a non-intercepting grid, in which case the beam could be turned on and off with little wasted energy. A DC supply of 850 kV would be frightening to behold. When it arcs, the problem of crowbaring the stored energy would be difficult indeed. Furthermore, I am not sure one could design an electron gun for an X-band size beam which would hold off 850 kV DC, at least not one which would be cost effective.

KLYSTRON DESIGN AND OPERATING PARAMETERS

At SLAC we faced the problem of going from 60 MW at S-band to 75 MW at X-band. How did we modify our klystron design and operating parameters to achieve this? We knew that we faced a formidable increase in gradients, about 4 to 1, both RF and DC, and in beam power density as the structure and beam size decreased in going to four times higher frequency. First, let's consider what would happen if we did a direct scaling of the klystron from S-band to X-band, maintaining voltages and currents. As Table 1 shows, the results are rather startling. Cathode current density goes from a comfortable 6.25 A/cm^2 to 100 A/cm^2. Any value over 10 A/cm^2 will greatly shorten tube life. In fact, no one knows how to get 100 A/cm^2 reliably. The maximum gun gradient goes from 175 kV/cm to 700 kV/cm. Anything over 350 kV/cm is known to cause unacceptable gun arcing. Magnetic focusing field increases from 1500 gauss to 6000 gauss, which rules out any low-cost, permanent magnet solution. The problems continue.

TABLE 1. Scaling of Klystron S-Band to X-Band

Frequency	S-Band	Direct Scale X-Band	Modified Scale X-Band
Vo (Kv)	350	350	490
Io (A)	414	414	257
μ perv (Io/Vo$^{3/2}$)	2.0	2.0	0.75
Cathode Current Density (A/cm^2)	6.25	100	7.5
Cathode Area Convergence	25	25	125
Maximum Gun Gradient	175	700	229
Focusing Field (Gauss)	1500	6000	3120
RF Power Out (MW)	60	60	75
*Efficiency %	41	41	60

*Eff ≈ 74 - 16x μ perv

Although not shown in the table, an increase by a factor of four in RF gradient in the penultimate and output cavities will ensure unacceptable RF pulse shortening while the four times increase in the RF window gradient will guarantee its failure.

How do we produce a reliable X-band tube at the 75 MW level? Starting with the electron gun, we do it by raising rather than lowering the beam voltage. This allowed us to decrease the beam current with a disproportionate decrease in the microperveance, Io/Vo$^{3/2}$ of the electron gun. The next column in the table shows what happens to the operating parameters when we go from 350 kV and 414 A to 490 kV and 257 A. Electron gun microperveance goes from 2.0 to 0.75 while the required solenoid focusing field is approximately cut in half.

We now select a cathode diameter which ensures acceptable cathode current density. We chose 7.5 A/cm^2. This compares with 6.25 A/cm^2 for the S-band tube. This requires an extraordinary increase in the area convergence of the electron gun, from 25 to 1 to about 125 to 1. But this is OK. Although a high area convergence is not achievable with a high microperveance like 2.0, it is relatively straightforward to achieve at a microperveance as low as 0.75 using today's gun design programs. Figure 3 shows a typical gun run for the 125 to 1 area convergence gun.

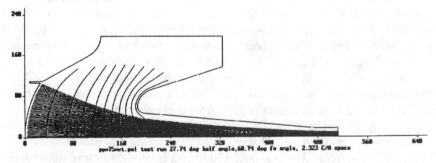

FIGURE 3. Simulation of .75 E Gun with 125/1 Beam Area Convergence

What happens to the performance of the tube? The power went from the 60 MW of the S-band tube, to 75 MW for the X-band tube, while efficiency went from a measured value of 41% for the S-band tube to a measured value of 60% for the X-band tube. This improvement in efficiency was predictable. The expression shown here (Eff = 74 - 16 microperv), though empirical, fits a wide range of klystrons which have been developed over the last quarter century. Note that even the gradient goes from an unacceptable 700 kV/cm to a very acceptable 229 kV/cm. The large increase in the cathode area increases the cathode/anode spacing. This is one of those rare instances when nature is kind. Increasing the voltage frees up the design, allowing a wide range of problems to be addressed.

We are not done with our design yet. We still have breakdown in the buncher cavities, the output cavity and the window, plus a still unacceptable solenoid power. Figure 4 is a cross-section of one of our X-band klystrons which will illustrate the design modifications which we used to solve the remaining problems.

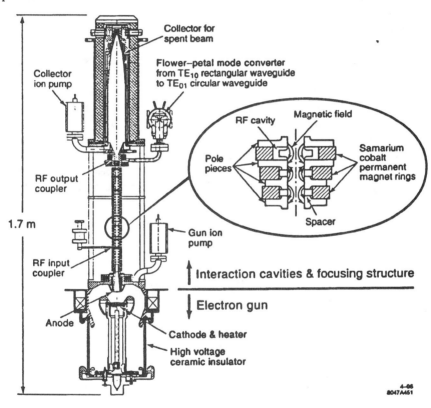

FIGURE 4. Cross Section of PPM-Focused 75 MW X-Band Klystron

To begin with, beam focusing was achieved using a periodic reversal of washer-shaped, high-energy rare earth magnets, which are split and attached to the tube after it is baked out. It is necessary only that the RMS value of this periodic magnetic field be made equal to the solenoidal value which is required to focus the same beam. This type of focusing is used on virtually all traveling-wave tubes. The difference between what we do and what is done on the traveling-wave tube, is that we use a magnet period which is very short compared to a plasma wavelength. This luxury is not available to the traveling-wave tube designer because of the unfavorable form factor of the TWT. We have 10 reversals of the field for every plasma wavelength. The plasma wavelength is, in a sense, a measure of the reaction distance of the electron beam to a perturbation. Hence, as long as perturbations are random, such as would be caused by the unavoidable tolerances and internal variations of commercial grade magnets, the effects are washed out in the big picture.

The short magnet period also ensures a very low stop-band voltage, that voltage below which no current is transmitted. This velocity filter characteristic is basic to all PPM-focused electron beams. We operate with a stop-band voltage of 3% of the operating voltage. This ensures that the small amount of current which is dropped out in the rise and fall of the beam pulse results in a negligible amount of heating of the klystron body. Finally, the use of a very short magnet period gives us a focusing field which is significantly lower on the axis than it is at the wall, hence the beam is trapped in a deep, potential well. As it bunches, it is less inclined to expand into the wall causing heating problems. To further reduce the chance of beam interception, we profile the field as shown in Figure 5, making it higher where bunching is greatest. The success of the conversion from solenoid to PPM-focusing is attested to by the results: More than 99.9% beam transmission to the collector with no RF, less than 1% beam power intercepted with full RF output.

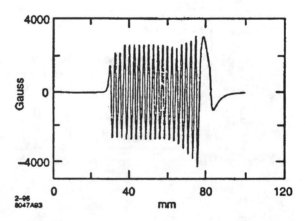

FIGURE 5. *Magnetic Field Profile for Best Efficiency Without Beam Interception*

Progressing down the tube, we solved the problem of the penultimate cavity gradient by replacing the single penultimate cavity with three cavities which share the voltage required to provide the final bunching of the beam. We used a similar trick in the design of the output circuit in which the single output cavity is replaced by a five-cell, disc-loaded, traveling-wave output circuit. The cartoon in Figure 6 shows the last of the three buncher cavities along with the disc-loaded output circuit and the tailpipe which leads to the collector.

The principle involved in reducing the output cavity gradient is the following: The voltage across the cavity has to be comparable with the beam voltage in order for the beam to be slowed down sufficiently to get good efficiency. A voltage of 490 kV across a single X-band cavity produces a maximum surface gradient well in excess of the breakdown value, which turns out to be around 1 MeV/cm. If, instead, one designs an output circuit having a number of cells, with the cells operating at approximately equal voltage, one can reduce the maximum gradient by a factor of at least two or three. To achieve the equal voltage per cell requires that the impedance from cell to cell be properly tapered. This is achieved by increasing the disc diameter from cell to cell. In order to obtain the best possible efficiency, there is also a taper in the phase velocity of the circuit from end-to-end so that the wave stays in synchronism with the slowing electron bunches.

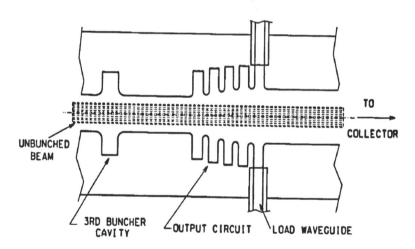

Figure 6. Cartoon of Final Buncher Cavity and 5-Cell Disc-Loaded Output Circuit.

The output window breakdown problem was solved by changing from the TE_{11} mode for the round disc window to the TE_{01} mode. Figure 7 illustrates the field lines for these modes. The TE_{01} mode, which is the well-known, low-loss mode, has the distinct advantage that there are no field lines terminating on the wall. Our TE_{11} window failures were all related to the field terminating on the wall, as evidenced by the fact that failed windows always showed a crack emanating along the ceramic, originating where the field terminates, never at 90° to that plane. The use of the TE_{01} mode was made practical by the development at SLAC of the flower petal TE_{10} to TE_{01} mode converter, which performs with over 99% efficiency and requires very little space. From the standpoint of the accelerator, use of a TE_{01} window in the klystron is a natural in that the pulse compressor also operates in the TE_{01} mode.

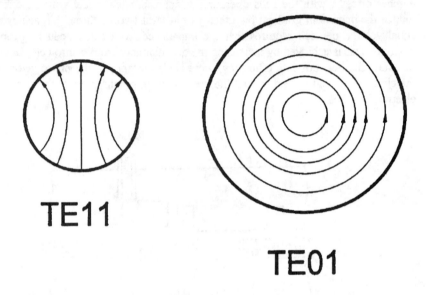

FIGURE 7. Comparison of Two Window Modes

We gained a further advantage in the window power handling capability by converting from a resonant half wavelength window to a traveling quarter-wave window. Figure 8 illustrates the field in the region of the window for the standing-wave case and the traveling-wave case. In the former, the standing wave is in the ceramic and there is a traveling wave on each side. This results in a maximum of the electric field at the surface of the window. In the latter, irises of appropriate dimensions are used to ensure that a match is achieved at the window, which results in a traveling wave passing through the window with standing waves on each side of the ceramic. In the latter case, the surface field is less than half of what it is for the standing-wave case. Assuming that the damage done in window breakdown is caused by surface field, the traveling-wave window is capable of supporting four times the power of the standing-wave window.

In practice, we often had failures in the TE_{11} standing-wave windows at as little as 20-MW RF power, while the TE_{01} mode traveling-wave window is good up to 100 MW. At this level we have had two failures in long-term testing in a X-band resonant ring. Each failure occurred at the point where electric field was highest. A puncture having the appearance of a volcano occurred in each case, suggesting that failure is initiated by a small occlusion of some sort in the ceramic. Our current efforts are in the direction of raising the margin by using improved hot, isostatic-pressed ceramics. We have never lost a hot, isostatic-pressed window in the resonant ring, but they are very difficult to metalize and braze.

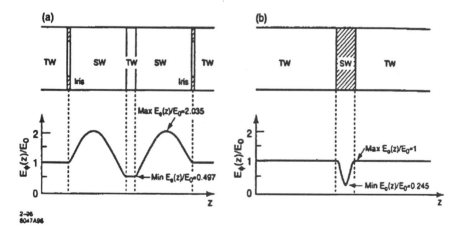

FIGURE 8. Comparison of the Reduced-Field TW Window (a) with a Half-Wave Resonant SW Window (b), both Operating in the TE_{01} Circular Mode. In Addition to the TW Version having Substantially Lower RF Electric Field at the Window Surface, the Integrated Dielectric Power Losses are only 23% those of the Half-Wave Resonant Window. The Horizontal Scale is Expanded for Clarity.

The design tricks which we used to "scale" the 60-MW S-band klystron, four times in frequency to high X-band, made it possible to keep the design and operating parameters within what are considered to be safe limits for long life and high reliability. Unfortunately, there is no way to get another factor of four in operating frequency from the klystron while meeting these same design and performance criteria. For instance, scaling the X-band klystron to 45 GHz, without increasing cathode current density, would require developing a reliable, long-life electron gun, having a beam area convergence of about 2000 to 1. Furthermore, the trick used for reducing the surface gradients in the buncher and output cavities would require 12 buncher cavities and a 20-cavity, traveling-wave output circuit, both of which appear to be impractical and not at all consistent with low cost. In fact, it is hard to imagine a 45 GHz, 75-MW peak-power amplifier of any known type, based on a low-cost design, which would provide long life.

Laser Acceleration in Vacuum

J.L. Hsu, T. Katsouleas
University of Southern California, Los Angeles, CA 90089-0484

W.B. Mori
University of California, Los Angeles, CA 90024

C. B. Schroeder, J.S. Wurtele
Lawrence Berkeley National Laboratory, Berkeley, CA 94720

Abstract. This paper explores the use of the large electric fields of high-brightness lasers (e.g., up to order TV/cm) to accelerate particles. Unfortunately, as is well known, it is difficult to couple the vacuum field of the laser to particles so as to achieve a net energy gain. In principle, the energy gain near the focus of the laser can be quite high, i.e., on the order of the work done in crossing the focus $\Delta\gamma = \sqrt{\pi} eEw \sim 30 MeV\sqrt{P/1TW}$, where P is the laser power. In order to retain this energy, the particles must be in the highly nonlinear regime ($V_{osc}/c >> 1$) or must be separated from the laser within a distance on the order of a Rayleigh length from the focus. In this work, we explore the acceleration and output energy distribution of an electron beam injected at various angles and injection energies into a focused laser beam. Insight into the physical mechanism of energy gain is obtained by separating the contributions from the longitudinal and transverse laser field components.

INTRODUCTION

The rapid development of high-brightness lasers leads us to re-examine the interaction of electron beams with vacuum focused lasers. For example, PetaWatt 1μm lasers are nearly available. For such lasers, the electric fields at focus will approach a TV/cm. There has been much considerable previous work on this topic, so the limitations to energy gain via a linear interaction are by now well known[1]. This paper is concerned with the highly non-linear regime (normalized quiver velocity $eE/m\omega c >> 1$) where a net energy gain is possible. Here, we present preliminary numerical results of the net energy gain, energy spread and angular spread that may be expected by injecting an electron beam at various angles into a focused PetaWatt-class laser. The goal of the simulation is to determine the net work done on a relativistic electron as it propagates through a laser focal zone. Insight into the physical mechanism responsible for the energy gain is obtained by separating the contribution from the longitudinal and transverse field components.

ALGORITHM

For a linearly polarized, cylindrically focused Gaussian beam with a vector potential in the +z direction, the field components are[2]

$$E_x = E_0 \frac{w_0}{w} \exp\left(-\frac{r^2}{w^2}\right) \cos\phi$$

$$E_z = 2E_0 \frac{w_0}{w} \frac{x}{kw^2} \exp\left(-\frac{r^2}{w^2}\right)\left(\sin\phi - \frac{z}{z_0}\cos\phi\right) \quad (1)$$

$$B_y = E_x$$

with the phase

$$\phi = kz - \omega t + \frac{r^2 z}{w^2 z_0} - \tan^{-1}\left(\frac{z}{z_0}\right) \quad (2)$$

where w_0 is the spot size, w is the laser field radius at position z, $z_0 = \pi w_0^2 / \lambda$ is the Rayleigh length, and k is the free-space wave number of the laser.

The movement of an electron in a combination of electric and magnetic fields is governed by the Lorentz force equation

$$\frac{d\vec{p}}{dt} = -e\left(\vec{E} + \vec{\beta} \times \vec{B}\right) \quad (3)$$

where $\vec{\beta} = \vec{v}/c$ for relativistic electrons. Substituting equation (1) into (3), the vector differential equation (3) becomes two scalar differential equations

$$\frac{dp_x}{dt} = -e(E_x - \beta_z B_y)$$

$$\frac{dp_z}{dt} = -e(E_z + \beta_x B_y) \quad (4)$$

The numerical algorithm is straightforward. The instantaneous EM fields acting on an electron are determined by giving an initial position of the electron. Given the initial velocity of this particle, then the EM force can push this electron into a new position with a new velocity after a small time step by using equations (4), and the

Time Center Leap Frog method[3].

The maximum energy gain from the acceleration[4] can be evaluated and cpmpared to analytic estimates.. The approximate value is given by estimating maximum work done by the laser field when the electron crosses the laser beam:

$$\Delta\gamma^{max} \approx \int_{-w}^{+w} \vec{E} \cdot d\vec{r}$$
$$\approx \sqrt{\pi} \cdot w_0 \cdot E_\perp + \int E_z dz \quad (5)$$
$$\approx 30 MeV \sqrt{\frac{P}{1TW}}$$

where P is the laser power in units of TW. For example, if we have a 100 TW laser, then the maximum energy from gain is no more than 300MeV. The last expression in (6) follows from just the transverse term. An interesting point is that the energy gain from the longitudinal $\int E_z dz$ term within a Rayleigh length z_0 scales identically with $w_0 \cdot E_\perp$, since $E_z \propto E_\perp \cdot \lambda / w_0$ from Gauss' law and $z_0 \propto w_0^2 / \lambda$. In the following section, we use the model just described to analyze three specific laser acceleration geometries: one is injecting electrons at a small incident angle; the second is coaxial injection; and the third is injecting down the axis of the two crossed laser beams.

SIMULATION RESULTS

I. Small Incident Angle Injection

In Fig.1, the main plot in the center shows a laser beam propagating from $z = -72000$ μm to $z = +72000$ μm in the z direction. An electron is injected from the left bottom corner where $z = -72000$ μm crossing the laser beam with a small incident angle θ. The small text boxes in the upper section show the parameters used in the example: the laser wavelength *lambda* is 10.6 μm, which corresponds to a CO_2 laser, in order to compare to parameters of UCLA CO_2 laser experiments; The laser waist width w is 200 μm; the normalized laser amplitude, $eE_0 / m\omega c \equiv aa$, is 0.4; the initial electron energy *gamma* is 32; and the electron incident angle *theta* is 0.04 rad. The energy of the electron γ is plotted versus the propagating direction z from -72000 μm to $+72000$ μm and shown in the small box in the left bottom side. We can see from the *gamma-z* plot that γ increases and decreases as the particle slips in phase behind the light wave, finally reaching the same value as its initial energy. Note that the full range of z in the inset is also from -72000 μm to $+72000$ μm. In this example, *aa* is not large enough to induce nonlinear effects, so no net energy gain results, as expected from the Lawson-Woodward theorem[1].

In the second example (Fig. 2), we increase *aa* from 0.4 to 4, which means that the laser power is 100 times that of the first one. From the inset γ-z plot in the left, we see that there is a net increase in the energy γ due to the interaction. The final energy at the point $z = 72000\mu m$ depends on the initial phase of the electron injection into the laser fields. The final energies of the injected electrons are plotted versus the initial laser phase in the lower left inset of Fig 3. This plot shows that for a particular initial laser phase, the maximum $\Delta\gamma$ is near 27.4; that is 5% of the maximum value we estimated using the work done by the laser in equation(5).

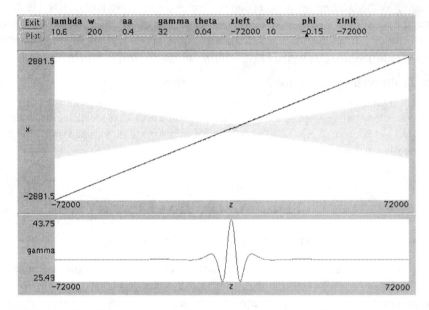

FIGURE 1. Particle trajectory and (below) particle energy (γ) vs. z for *aa*=0.4.

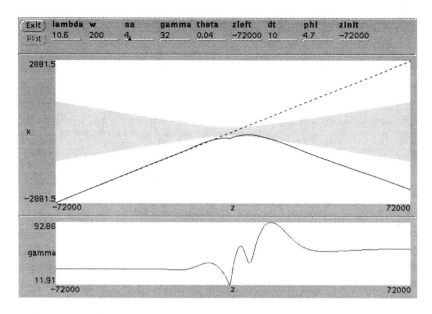

FIGURE 2. Particle trajectory and (below) particle energy gain vs. *z* for *aa* = 4. γ increase from 32 to 59.4(dotted line shows the incident direction for this electron).

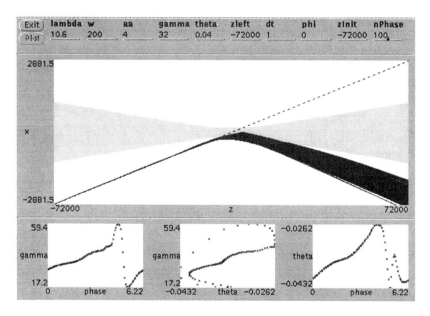

FIGURE 3. Particle trajectories for 100 initial laser phases from 0 to 2π and insets: final γ vs. initial laser phase, finial γ ys. outgoing angle θ, θ vs. laser phase.

II. Coaxial Injection

Before considering the coaxial injection into focused laser beams, consider the movement of electrons in infinite plane waves. The exact solution for an electron moving in an infinite plane wave can be expressed as a drifting "figure eight" illustrated in Fig.4. Adding the drift to the figure eight gives the trajectory on the right in Fig.4. Using a large beam waist w in our simulation program approximates the plane wave limit; the trajectory of an electron in Fig.5 exhibits similar behavior to the exact solution in an infinite plane wave as shown in Fig.5.

When the laser beam is focused to small spot size w, the trajectories of coaxially injected electrons are quite different. The trajectories of coaxially injected electrons with initial laser phases distributed from 0 to 2π are shown in Fig.6. These electrons are scattered before they reach the midpoint. The γ-*theta* plot shows the final γ measured at the point $z = 72000$ μm versus different outgoing angle *theta*. As the first inset (γ-phase plot) shows, at the injection phases of 0 and π, the final energy is relatively insensitive to injection phases. Therefore, particle injection at these phases will result in better output beam quality. Notice the lower left inset of Fig.3 shows that for small angle scattering, the final energy is very sensitive to injection phase for phases near the phase of about the peak energy gain. Therefore we expect better beam quality from coaxial injection at carefully chosen injection phases (0 or π) than from small angle injection. The maximum energy gain in the coaxial example is 3.8% of the maximum value from equation (5). Increasing the initial γ from 32 to 64 made the electrons go through the beam focus (Fig.7); in that case the maximum gain was 9.1% of equation (5).

In order to separate the effects of transverse and longitudinal fields in the acceleration process, we turn off the longitudinal field and find the interaction with the transverse field only (Fig.8). We observe that the electron trajectories are considerably altered by the presence of the E_z field, even though we found that the $\int E_z dz$ contribution to net energy gain was much smaller than the $\int E_\perp dy$ term for our examples.

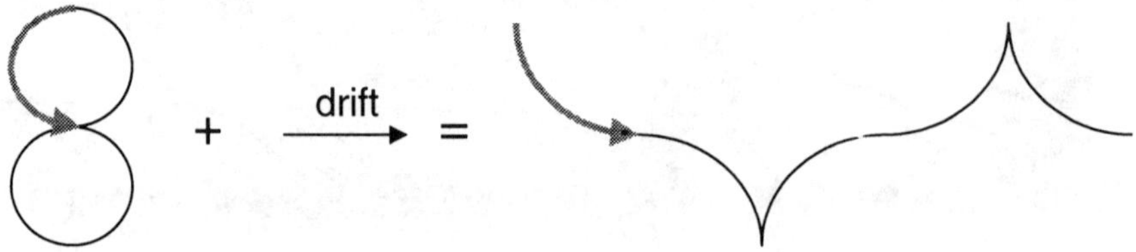

FIGURE 4. Electron movement in infinite plane wave

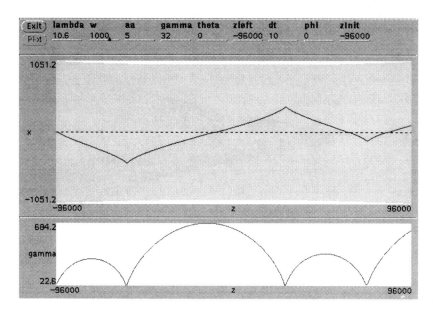

FIGURE 5. Particle trajectory and (below) γ–z plot for coaxial injection with large laser waist

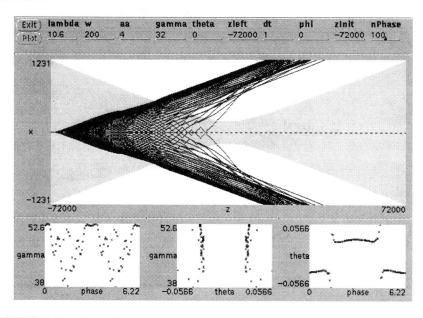

FIGURE 6. Particle trajectories and final energies and insets same as Fig 2.

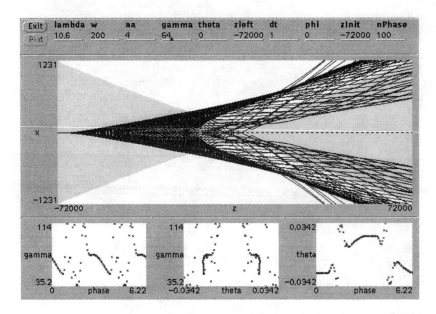

FIGURE 7. Same as Fig. 6 except initial $\gamma = 64$.

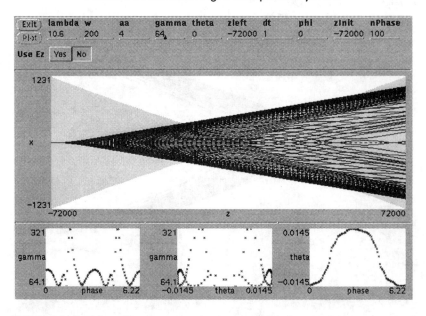

FIGURE 8. Same as Fig. 7 except the longitudinal electric field is turned off.

III. Crossed Beam Injection

A laser geometry that has received attention is the case of two gaussian laser beams crossed at a small angle with respect to the axis of electron injection[5,6]. The advantage of this geometry is that the transverse electric and magnetic fields will cancel on axis, leaving only the axial electric field to accelerate the particles. Fig.9 shows the simulation results of electrons injected at various initial laser phases on axis into the crossed beams which are positioned at an angle of 0.04 rads with respect to the electron beam. As the inset shows, there was no net energy gain for any initial laser phases. This result can be explained by realizing that by arranging the geometry such that there is only an electric field along the direction of motion, we have eliminated the nonlinear forces which are necessary for laser acceleration in vacuum. This is consistent with the results of P. Sprangle et al.[7]

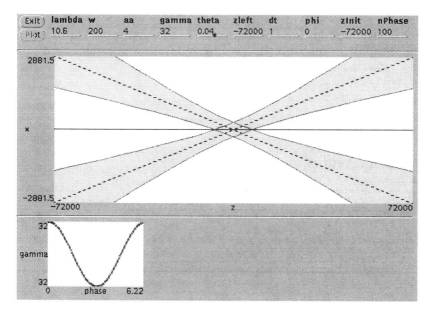

FIGURE 9 Numerical simulation showing no energy gain for any injection phase in two crossed lasers (dotted lines show laser beam axes).

CONCLUSION

The simulation results show that net energy gain can be extracted from a single laser via nonlinear interactions. The nonlinear energy gain comes from the transverse fields; while the longitudinal field affects the path of electrons but does not increase their energy gain. If initial injection energy and angle and laser amplitude are chosen properly, large scattering spread angle can be avoided. For further research, characteristics of the outgoing electron beam with respect to the injected beam emittance need to be investigated. Other acceleration schemes, such as axicon focused lasers[5], or standing wave acceleration in two counterpropagating lasers[8] are also of interest for further simulations with this code.

ACKNOWLEDGMENTS

Work supported by US Department of Energy DOE - AC#DE - FG03 - 92ER40745. The assistance of R. Kinter and D.Gordon in coding the numerical model is gratefully acknowledged.

REFERECES

[1] P.M. Woodward, *JIEE 93*, 1947, p.1554.

[2] R. Guenther, *Modern Optics,* (Wiley, New York 1990), p.336.

[3] C.K. Birdsall and A.B. Langdon, *Plasma Physics via Computer Simulation*, (McGraw-Hill, NY 1985).

[4] W. Mori and T. Katsouleas, in *Advanced Accelerator Concepts.* AIP Conf. Proc. **No. 335**, (AIP, NY 1994), p.112.

[5] L. Steinhauer et al, in *Advanced Accelerator Concepts,* AIP Conf. Proc. **No. 335**, (AIP, NY 1994), p.131.

[6] C. M. Haaland, Optics Comm. **114**, 280 (1995).

[7] P. Sprangle, Optics Comm. **124**, 69 (1996).

[8] C. Barty, private communication.

New Advances in Inverse Cerenkov Acceleration

W. D. Kimura,[*] M. Babzien,[†] D. B. Cline,[‡] R. B. Fiorito,[††]
J. R. Fontana,[‡‡] J. C. Gallardo,[†] S. C. Gottschalk,[*] K. P. Kusche,[*†]
Y. Liu,[‡] I. V. Pogorelsky,[†] D. C. Quimby,[*] R. H. Pantell,[§]
D. W. Rule,[††] J. Skaritka,[†] J. Sandweiss,[§§] A. van Steenbergen,[†] and
V. Yakimenko[†]

[*]STI Optronics, Inc. Bellevue, WA 98004-1495
[†]Brookhaven National Laboratory, Upton, NY 11973-5000
[‡]University of California at Los Angeles, Los Angeles, CA 90024-1594
[††]Naval Surface Warfare Center, West Bethesda, MD 20817-5700
[‡‡]University of California, Santa Barbara, CA 93106
[§]Stanford University, Stanford, CA 94305
[§§]Yale University, New Haven, CT 06511-8167

Abstract. Inverse Cerenkov acceleration (ICA) has entered a new phase in its development. The issue of staging and rephasing the optical wave with a microbunched electron beam is now being examined. This ability to accelerate over multiple stages is important for scaling laser accelerator devices to higher energies. An inverse free electron laser (IFEL) will be positioned upstream from the ICA experiment and used to prebunch the electrons. These electrons will then be focused into the ICA interaction region for rephasing and acceleration by the laser beam. Issues that will be examined during these combined ICA/IFEL experiments include rephasing the laser beam with the microbunches, minimizing bunch smearing, and trapping the electrons in an acceleration bucket.

INTRODUCTION

Scaling laser particle accelerators to higher energies will most likely require using multiple acceleration stages in which the optical wave must be properly rephased with the microbunched electron beam at the optimum acceleration point within each stage. This situation is analogous to bunching and trapping in conventional microwave linacs and is important for maximizing the efficiency of the acceleration process. However, as explained later, these microbunched beams consist of electron bunches with longitudinal density distributions a fraction of an optical wave in length separated by the laser wavelength. These bunch dimensions are orders of magnitude smaller than conventional microwave linacs, which gives rise to new issues that must be addressed.

As part of its effort to demonstrate 100-MeV net acceleration (1), the inverse Cerenkov acceleration (ICA) experiment is entering into a new phase whose aim is to examine this issue of staging and accelerating a microbunched beam. The ICA experiment is located on the Accelerator Test Facility (ATF) at Brookhaven National Laboratory (BNL), where laser acceleration of the 50-MeV ATF electron-beam (e-beam) has been routinely obtained since the experiment first began running in 1995 (2).

In ICA, the e-beam interacts with the laser beam inside a cell filled with H_2 gas (3). The index of refraction provided by the gas slows the phase velocity of the light wave such that the electrons stay in phase with the light wave over the entire interaction length. This phase matching condition is satisfied when the laser beam crosses the e-beam at the Cerenkov angle defined by $\theta_c = \cos^{-1}(1/n\beta)$, where n is the refractive index of the gas and β is the electron velocity divided by the velocity of light.

The electron pulse length entering the ICA gas cell is ~10 psec long. A 20-GW CO_2 laser ($\lambda = 10.6$ μm), located at the ATF, drives the ICA process and has a pulse length of ~200 psec. Hence, during the ICA interaction the electrons within the 10-psec pulse experience all phases of the light wave. In other words, some electrons are accelerated, some are decelerated, and some experience little energy change. This energy modulation of the e-beam results in bunching of electrons. After an appropriate drift distance, the electrons' longitudinal distribution will have density variations characterized by the formation of electron microbunches. Our model simulations indicate that these microbunches will have bunch lengths of order 2 μm with the bunches separated by the laser wavelength, i.e., 10.6 μm. Thus, within the 10-psec electron pulse there will now exist a train of 7-fsec microbunches separated by 35 fsec.

This same microbunching process occurs with many of the other laser acceleration schemes. In particular, the process is nearly identical for an inverse free electron laser (IFEL). Such an IFEL, in fact, has already demonstrated laser acceleration at the ATF (4).

At the 1996 ICA Workshop (5), R. H. Pantell recognized that an IFEL would perform better as a prebuncher than an ICA prebuncher. This is because during the ICA process, the electrons experience some scattering off the gas molecules during the interaction. For high energy e-beams (i.e., high γ) this is not a serious issue; however, for the relatively low 50-MeV ATF beam, this is critical especially during prebunching where slight changes in the direction of the electrons within the interaction region can interfere with the bunching process. An IFEL, of course, operates in a vacuum and avoids this scattering problem.

Therefore, an important new development in this program is the joining of the ICA experiment with the BNL IFEL experiment headed by A. van Steenbergen.

The BNL IFEL will be used as the prebuncher for the ICA experiment. The long term goal of the ICA experiment is to use the upgraded ATF CO_2 laser (6), which will be capable of 1 TW peak power, to accelerate the microbunched electrons from the IFEL and demonstrate 100-MeV net energy gain. This combined ICA/IFEL experiment is described in more detail next.

COMBINED ICA/IFEL EXPERIMENT

Figure 1 shows a conceptual layout for the 100-MeV laser accelerator. The electrons enter the IFEL prebuncher from the left on the drawing. The drive laser beam is split into two beams. A small portion of the beam (e.g., ~10 MW) is sent into the 0.5-m long wiggler of the IFEL prebuncher. Most of the laser power, e.g., ~250 GW, is sent to the ICA laser accelerator. (Based upon model simulations, a minimum of 250 GW is needed to demonstrate 100-MeV energy gain. Although, as mentioned, the final upgraded ATF CO_2 laser will be capable of ~1 TW output.)

Downstream of the IFEL will be a drift space of about 2-m length and quadrupole magnets. The former is needed to allow the electrons to microbunch after leaving the IFEL.[1] The latter are needed to focus the electrons into the ICA gas cell since the e-beam is diverging when it exits the IFEL. Thus, the position of the ICA gas cell relative to the IFEL depends upon the optimum bunching distance for the microbunch formation. This distance is a function of the e-beam energy and the amount of energy modulation imparted by the laser beam within the IFEL. Of course, the distance between the IFEL and ICA gas cell is a fixed parameter. Therefore, for a given e-beam energy it is possible to have the electrons achieve maximum bunching density at this fixed distance by adjusting the amount of laser power delivered to the IFEL.

The ICA gas cell will be similar in design to the present cell. A plan view is shown in Fig. 2. The laser enters the cell from the top of the drawing. It reflects off the 45° mirror towards an axicon mirror on the right. The axicon mirror focuses the laser beam at the Cerenkov angle onto the e-beam, which is entering the cell from the right in the drawing. The line focus created by the axicon is approximately 20-cm long in the present experiment and represents the interaction region (see Fig. 2) for the ICA process. For the 100-MeV demonstration experiment, this length must be increased to ~30 cm. This will require fabricating a larger gas cell. However, in the near-term, the existing gas cell will be used.

---

[1] A magnetic chicane could be used after the IFEL to shorten the bunching distance. This option is not being pursued because of the need to use quadrupoles to focus the e-beam into the ICA gas cell. These quadrupoles must occupy a finite drift space, thereby defeating the purpose of using the magnetic chicane.

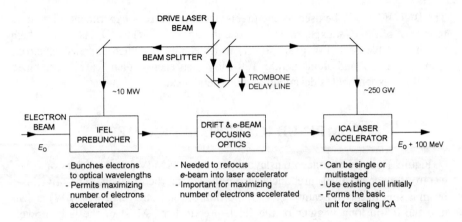

FIGURE 1. Conceptual layout for 100-MeV ICA/IFEL laser acceleration demonstration experiment.

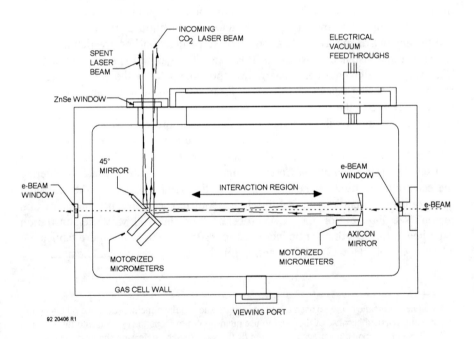

FIGURE 2. Plan view of presently used ICA gas cell.

The electrons enter and exit the gas cell through 2.1-μm thick diamond e-beam windows. They travel through a 0.5 mm diameter hole in the axicon mirror and a 1 mm diameter hole in the 45° mirror.

The Cerenkov angle during the present experiments is 20 mrad, which at STP conditions, implies the need for a hydrogen gas pressure of ~1.8 atm to provide the proper index of refraction for phase matching. While the optimum Cerenkov angle has not been determined yet for the 100-MeV demonstration experiment, it is highly probable that a comparable or slightly larger angle will be used. This is because smaller angles reduce the acceleration gradient, while larger angles require higher gas pressures. The 2.1-μm thick diamond windows have withstood 3-atm pressure tests. Hence, higher gas pressures are possible and, therefore, slightly larger Cerenkov angles may be used.

Besides increasing the interaction length for the 100-MeV demonstration experiment, the optics inside the gas cell will also be modified to help them withstand the larger power from the upgraded ATF CO_2 laser. One possible way to increase the damage threshold of the optics is to use glancing incidence mirrors. For example, instead of using a 45° mirror, a higher angle of incidence mirror, say, 80°, can be used.

The issue of gas breakdown at the higher laser powers has been theoretically examined (7). This breakdown threshold increases with decreasing laser pulse length. The upgraded ATF CO_2 laser will feature the capability of generating pulse lengths less than its present 200 psec. In fact, this pulse shortening is important for achieving the 1 TW goal. Hence, we anticipate the shorter pulse length of the upgraded laser will help alleviate gas breakdown problems.

Even if gas breakdown does occurs, it is not clear that this will necessarily interfere with the ICA process. We have already shown analytically (8) that a plasma within the interaction region will not appreciably disturb the Cerenkov phase matching condition.[2] The primary issue will be potential absorption or reflection of the laser light by the plasma. This situation can be avoided in principle by utilizing shorter wavelength laser light where the gas breakdown plasma density can be lower than the critical density (9). However, for the infrared wavelength of the CO_2 laser, there are less options available to avoid absorption by the laser light. Fortunately, gas breakdown requires a finite time to occur. Hence, by adjusting the arrival of the microbunches to occur at the beginning of the laser pulse, it may be possible to accelerate the electrons before appreciable laser energy is absorbed by any breakdown.

---

[2] The laser beam pathlength through the plasma is too short to significantly alter the phase of the light wave. In addition, the Cerenkov angle for phase matching changes in the same direction as the refraction of the light wave through the plasma, thereby automatically compensating for this change in angle.

EXPERIMENTAL PLANS

A detailed design for the combined ICA/IFEL experiment is being developed. Issues that are being examined include:

1) Design of the beamline. The position of the IFEL relative to the ICA cell is only one of the critical design issues that must be considered. The positions and lengths of the focusing quadrupole magnets are also important, especially for the quadrupoles between the IFEL and the ICA cell. This is because electrons being focused by the quadrupoles can travel different pathlengths before they reach the ICA cell. This can lead to bunch smearing, which becomes a potentially serious issue because the bunch lengths are only microns long. Therefore, pathlengths must be controlled to within microns.

2) Quality of the incoming e-beam. The intrinsic energy spread and emittance of the e-beam affect the quality of the bunching. The aforementioned bunch smearing can be minimized by reducing the e-beam emittance. Fortunately, the ATF e-beam has recently demonstrated normalized rms emittances as low as 1 π-mm-mrad (7).

3) Design of the IFEL wiggler. One attractive feature of the BNL IFEL wiggler is the ease with which it can be altered. This may be important because preliminary analysis indicates that the optimum IFEL wiggler configuration is one using a uniform rather than tapered wiggler design. The present wiggler uses a tapered design; thus, it may be necessary to change the wiggler configuration for the combined ICA/IFEL experiment.

4) Rephasing of the laser beam with the e-beam. A key issue to be examined during the combined experiment is controlling the phase of the laser light in the ICA gas cell relative to the microbunches produced by the IFEL. The phase of the microbunches is controlled by the phase of the laser beam entering the IFEL. Hence, controlling the phase between the laser beam within the ICA cell and the microbunches is equivalent to controlling the phase between the two laser beams sent to the IFEL and ICA devices (see Fig. 1). We anticipate that the optical transport line for the laser beam directed to the ICA cell will include a "trombone" delay line with a pair of mirrors translated by standard piezoelectric pushers capable of submicron accuracy (see Fig. 1). Optimum phase will be

determined during the experiment by scanning the piezoelectric pushers over a 10.6 μm distance and observing the energy of the microbunches exiting the ICA cell until maximum gain is observed. One experimental issue is the stability of the optical pathlengths to the IFEL and ICA devices. These lengths must be stable to ~1 μm or better if the microbunch lengths are ~2 μm.[3] To ensure this stability it may be necessary to utilize an active stabilization system whereby the piezoelectric pushers are driven by a feedback loop coupled to a detector that senses changes in pathlengths. Although such systems have been made, the complexity of the ICA optical system makes applying this technique a challenge.

5) Detecting the microbunches. Detecting and characterizing 2-μm long (7-fsec) microbunches is a nontrivial task. Currently available e-beam diagnostics, such as those using streak cameras, cannot resolve this short of pulse lengths. It is important to measure the microbunch characteristics in order to compare with theory and understand how well the rephasing and trapping process is occurring. One promising technique is to use coherent transition radiation (CTR) (10). The microbunches generate CTR at harmonics of the laser wavelength. By measuring the relative strengths of these harmonic components it is possible to determine the degree of microbunching (i.e., the relative amount of electrons in the microbunches). Harmonic radiation was recently detected from microbunches produced by the BNL IFEL (11). It is not clear whether this technique can reveal information about the microbunch length. We are currently exploring other diagnostic ideas to address this issue.

6) Permitting IFEL only experiments. The IFEL project has plans to test a second wiggler module that is fed by the first one. To enable this on the same beamline, we are designing the system so that the ICA gas cell can be exchanged with the second IFEL module. It may be necessary to also change the quadrupoles after the first IFEL module to provide optimum e-beam focusing into the second IFEL module.

---

[3] The 1-μm stability value is a rough estimate. Acceleration within the 10.6-μm laser wavelength occurs within a half-wavelength interval (i.e., 5.3 μm) of the light wave. A 2-μm long microbunch occupies a significant portion of this half-wavelength. Therefore, for this microbunch to stay within this acceleration interval, its position should be stable to approximately ±1 μm.

7) Integrated ICA/IFEL model development. We are upgrading our existing ICA Monte Carlo code (12) to include an IFEL prebuncher. Electron beam focusing between the IFEL and the ICA cell is modeled using the TRANSPORT code. This ICA/IFEL code will be used to predict the results of the combined experiment and to optimize its design.

8) Space charge effects. Space charge can become an important issue when dealing with the very small dimensions of the microbunches. Preliminary analysis using PARMELA (13) indicates for the ATF conditions that space charge effects will not be a problem. Nonetheless, further PARMELA analysis is being performed to fully understand the scope of this effect and the conditions when it becomes a serious issue.

The next steps in our experimental plans are to move the ICA and IFEL hardware into their new positions for the combined experiment, as well as install additional beamline components, such as quadrupoles. The first experiments will be to verify nominal operation of each individual system separately. Then, the first combined experiments will be to send the prebunched beam from the IFEL into the ICA cell while delivering ~10 GW of laser power to the ICA cell. This lower laser power will avoid any potential optical damage issues, but still allows the rephasing and trapping measurements to occur. Once we have verified that the rephasing and acceleration processes are functioning properly, then the laser power will be increased entering the gas cell. This, of course, may require installation of the larger gas cell.

Several diagnostics will be used during the experiments. An energy spectrometer at the end of the beamline provides information about the e-beam energy distribution. As mentioned, CTR diagnostics may provide information about the microbunching characteristics. Other diagnostics provide information regarding the e-beam current and profile.

CLOSING

The joining of the IFEL with the ICA experiment represents an important new development in this research effort. This collaboration between these two experiments will be one of the first to examine the important issues of staging, rephasing, and trapping of the microbunches produced by laser accelerators. This combined experiment is currently being designed and assembled. First, experiments may occur in late 1997 or early 1998.

ACKNOWLEDGEMENTS

The authors would like to acknowledge the help and advice provided by Dr. I. Ben-Zvi, Dr. X. J. Wang, Dr. R. C. Fernow, and Dr. L. C. Steinhauer. This work was supported by the U.S. Department of Energy, Grant Nos. DE-FG06-93ER40803 and DE-AC02-76CH00016.

REFERENCES

1. Romea, R. D., Kimura, W. D., and Steinhauer, L. C., "100-MeV Inverse Cerenkov Accelerator Demonstration Experiment," in *Advanced Accelerator Concepts*, Fontana, WI, AIP Conference Proceedings No. 335, P. Schoessow, Ed., New York: American Institute of Physics, 1995, pp. 390–404.
2. Kimura, W. D., Kim, G. H., Romea, R. D., Steinhauer, L. C., Pogorelsky, I. V., Kusche, K. P., Fernow, R. C., Wang, X., and Liu, Y., *Phys. Rev. Lett.* **74**, 546–549 (1995)
3. Fontana, J. R. and Pantell, R. H., *J. Appl. Phys.* **54**, 4285–4288 (1983).
4. van Steenbergen, A., Gallardo, J., Sandweiss, J., Fang, J.-M., Babzien, M., Qiu, X., Skaritka, J., and Wang, X.-J., *Phys. Rev. Lett.* **77**, 2690–2693 (1996).
5. 1996 Inverse Cerenkov Acceleration Workshop, Feb. 15-17, 1997, Bellevue, Washington (unpublished).
6. Pogorelsky, I. V., Kimura, W. D., Fisher, C., Kannari, F., and Kurnit, N. A., "Approach to Compact Terawatt CO_2 Laser System for Particle Acceleration," in *Advanced Accelerator Concepts*, Fontana, WI, AIP Conference Proceedings No. 335, P. Schoessow, Ed., New York: American Institute of Physics, 1995, pp. 405–418.
7. Liu, Y., Cline, D., Pogorelsky, I. V., and Kimura, W. D., in preparation.
8. Steinhauer, L. C. and Kimura, W. D., *J. Appl. Phys.* **68**, 4929–4936 (1990).
9. Kimura, W. D., Romea, R. D., and Steinhauer, L. C., *Particle World* **4** (3), 22–31 (1995).
10. Rosenzweig, J., Travish, G., Tremaine, A., *Nucl. Inst. Meth. Phys. Res. A* **365**, 255–259 (1995).
11. Liu, Y. and Wang, X. J., Brookhaven National Laboratory, private communication.
12. Romea, R. D. and Kimura, W. D., *Phys. Rev. D* **42**, 1807–1818 (1990).
13. Bogacz, S. A., Fermi National Accelerator Laboratory, private communication.

Beam Driven Plasma Wake-Field Acceleration

Alexander N. Skrinsky

Budker Institute of Nuclear Physics, Novosibirsk 630090, Russia

Abstract. General problems of high gradient plasma wake-field acceleration are discussed. The mechanism of electric field excitation in a plasma by ultrarelativistic electron driver as well as a long-term evolution of the driver itself are qualitatively described. The idea of sequential acceleration based on a helical "delay-line" for driver beam train is proposed. It is shown that on the part of physics there are no prohibitions to build a full scale plasma wake-field accelerator of TeV energy range.

INP entered the field quite recently (4 years ago), and I shall present our steps made, only.

Suggestions and attempts on higher gradients were made by many people during decades and many times. Since 1960s, for us, at Novosibirsk, the main drive in the direction was *Linear Collider* hope for hundreds of GeV electron-positron collisions.

When we first presented publicly (1978) the real VLEPP project, which incorporates 100 MeV/m accelerating gradient for normal conducting short pulse linacs, this "high gradients problem" was considered by the majority as the main one in linear colliders (even obstacle!).

At that time, additionally to VLEPP project, we discussed the possibility to use the huge energy stored in existing and - especially - planned proton beams of super-accelerators. This option was called the *"proton klystron"*, and the way was shown how to excite by such beams 1 cm wave length range linear accelerating structures up to the maximum accelerating gradients limited by breakdown.

Since that time, in our Lab and in many other labs this level was proved practical for GHz (cm wave) range, short pulse, normal conducting accelerator structures - just having enough RF power and using proper technology for structures!

But it was (more or less) evident from the beginning that 100 MeV/m is close to the ultimate limit for metallic accelerating structures: the electric field at the surface is additionally several times higher and it starts producing "cold currents", and occasional discharges degrade the surface instead of improving it (training process fails). Hence, to reach much higher gradients we need to shift from the solid materials shaping of electromagnetic fields to plasma based structures. (At really high electric fields, the exposed surface in any case converts into something close to plasma.)

Thus we conclude:
- Plasma based devices is the "only" way to 0.3 Gev/m and higher!
- High energy ("rigid", more than 1 GeV) driver is the only way to the high gradient plasma accelerator (neither lasers nor low energy beams): we need "long" and high precision induced plasma field structure (in space and in time) - *of precise metallic surfaces level!* We ought to overcome plasma and beam-plasma (local) instabilities.
- Plasma should be arranged in advance - driving beam ionization is too weak. And ionization should be close to 100%.

Why plasma ?

First of all - the simplest question: why do we need plasma for efficient transfer of driving beam energy to the witness beam under acceleration?
- The longitudinal electric field - the <u>only</u> field which <u>accelerates</u> - is the same both in the beam frame and in the lab frame. In the beam frame, longitudinal field of the very thin and long driving bunch is much smaller than its transversal field. In the lab frame, transversal electric field is additionally γ times higher. Hence, direct use of a relativistic driving bunch for acceleration is extremely inefficient.
- Driving bunch traveling through the plasma excites oscillations of plasma electrons via its high <u>transversal</u> electric field; thus excited longitudinal electric field can be of the scale of the driver transversal field.
- Additionally, we can store excitation from many driving bunches.

Inside the plasma, plasma electrons and electromagnetic fields oscillate coherently, and more or less universal limit (if we intend to use efficient "resonant excitation") can be evaluated as follows: *electric field energy density should be less than energy density of plasma electrons.*

$$\frac{E_{ultim}^2}{4\pi} = m_e c^2 n_e$$

Hence,

$$E_{ultim} = \sqrt{4\pi \cdot n_e m c^2} = \sqrt{\frac{4\pi \cdot e^2}{r_e} \cdot n_e}$$

The same estimate can be obtained from Poisson equation

$$\text{div } \mathbf{E} = 4\pi\rho$$

by putting $\rho = e n_e$ and $\text{div } \mathbf{E} \sim E\omega_e/c$ there (ω_e is the plasma frequency).

The real accelerating field should be, say, 3 times lower. Hence, the plasma density needed to achieve the accelerating field E_{acc} should be not less than

$$n_e = \frac{9}{4\pi} \cdot \frac{r_e}{e^2} \cdot E_{acc}^2$$

To achieve 1 GeV/m, we need to use plasma density 10^{15} cm^{-3}.

We should try to excite non-propagating axially symmetric plasma oscillations. The obvious example is a plasma cylinder with longitudinally homogenous excited radial motion of plasma electrons. In such a system there is a radial electric field only, and the magnetic field is zero. For necessary high fields, short process duration and modest densities, we can assume:

plasma temperature - zero,
kinetic pressure - zero,
plasma without collisions,
only plasma electrons are mobile.

With this assumptions we can see that the electric field and plasma electrons oscillate coherently, and there is no influence of internal plasma oscillation on the outer plasma. Moreover, the excited plasma cylinder does not grow in time! (The excitation energy does not flow away and does not dissipate.)

But in such excitation structure there is no longitudinal electric field and, hence, no acceleration. To make acceleration possible, we have to excite the plasma cylinder with local phases shifted properly along it.

For beam driven excitation, the phase velocity along the axis is determined by the velocity of the driving beam; in all our considerations it will be the light velocity "c" since all the beams are ultra-relativistic. (Let us not forget - the driver current excites compensation plasma current, also)

Because of zero plasma temperature and no plasma collisions, plasma electrons do not enter continuos acceleration (no Landau damping, no plasma breakdown).

Local resonant frequency of plasma oscillations is the "electron plasma frequency":

$$\omega_e = \sqrt{\frac{4\pi e^2 n_e}{m_e}}$$

Optimal driver is a train of micro-bunches resonantly spaced by

$$\lambda_{pl} = \sqrt{\frac{\pi}{r_e n_e}}$$

<u>The driver required</u>

Let us evaluate the number of particles in the driver that are required to excite in plasma the accelerating field needed. We assume the driving beam radius to be about the minimal localization radius of plasma oscillations, which is
$$\lambda_{pl}/2\pi.$$
The balance of energies - the energy stored in "plasma tube" per 1 cm and the energy lost by driver particles at this distance is
$$\frac{E^2}{8\pi} \cdot \pi \cdot \frac{\lambda^2}{(2\pi)^2} \cdot 2 = e\, N_{need}\, E \cdot \frac{1}{2}.$$
Hence,
$$N_{need} = \frac{1}{8\pi} \cdot \frac{E\lambda^2}{e}.$$
For 1 GeV/m=30 kGs and 1 mm wave length, $N_{need} = 1 \cdot 10^{10}$ (of course, the estimate assumes all the particles are passing in a proper oscillation phase!).

To minimize the peak driver current, the driver particles should be distributed among the micro-bunches separated by distance λ. This train, while traveling at speed of light, should excite plasma oscillations resonantly. Hence, for plasma density $1 \cdot 10^{15}$ cm^{-3} micro-bunch spacing should be close to λ_{pl}=1mm.

If the driver particles are distributed in a 10 micro-bunch train of length = 1 cm, the mean current in the train is
$$\frac{1 \cdot 10^{10} \cdot 3 \cdot 10^{10}}{6 \cdot 10^{18}} = 50\,\text{A}.$$
The necessary peak current in micro-bunches should be 10 to 15 times higher. This estimate is correct for linear regime, when accelerating field is much lower than the ultimate one; if it comes closer - the efficiency is going down.

The effective computer codes named NOVOCODE and LCODE to analyze any driver-witness arrangements were developed (and under continuos improvement). Their main features are:
  axial symmetry assumption;
  characteristic times (lengths) hierarchy used:
    plasma oscillations 0.15 mm,
    (de)focusing length 10 cm - 1 m,
    deceleration length 10 - 100 m;
  zero plasma temperature;
  no plasma collisions;
  plasma electrons are treated hydro-dynamically;
  beams are treated in (axially symmetric) macro-particles approximation.

Using the estimations and these codes, we realized:
- The most efficient approach is to use the driving beam radius of $\lambda_{pl}/2\pi$.

- To minimize the driver peak density, the train of micro-bunches (around 10) should be used.
- Amplitude of transversal forces acting on driver and witness particles in excited plasma channel are much stronger than any external focusing. Hence, a care should be taken to position all the micro-bunches properly: the driver to be decelerated, the witness to be accelerated, and both of them to be in focusing phases during all the acceleration time!
- Micro-bunches should be longitudinally positioned not equidistantly (closer as the gradient grows) to remain stable.
- To direct leading micro-bunch properly and to prevent its radius from growing (still "no" plasma focusing fields induced!) we need strong and precise quadrupole focusing of the driving beam. (External focusing is necessary and important!)
- Axisymmetric "static" instability (over-focusing effects) is found to be dangerous at extreme gradients. A way to solve the problem is the proper matching of local micro-bunch emittance and local plasma focusing - to keep the driving beam radius constant and optimal.
- Axisymmetric resonant instability is found to be completely suppressed because of fast enough change in local driver particles frequency due to deceleration.
- Dipole instabilities are still not analyzed carefully - the non-axisymmetric code is in preparation, only. The hope is - these instabilities would be not dangerous - again, because coherent transversal oscillation frequencies have extremely high gradients along separate micro-bunches and between micro-bunches. The higher transversal modes are expected weak enough to be neglected completely.
- Energy losses of extremely narrow witness beam should be considered separately since the hydrodynamic approximation fails in the vicinity of the witness. By estimations, it was shown that the additional losses are not too detrimental (even for parameters of the high luminosity linear collider - less than 100 MeV/m).
- The proper use of strong plasma focusing (amplitude of transversal plasma field is about 1/3 of longitudinal plasma field, hence around 10 kGs; corresponding (de)focusing gradients up to 700 kGs/cm!) allows to keep witness emittance very low, in spite of multiple scattering in plasma - crucially important for high luminosity linear collider.
- Plasma longitudinal homogeneity and stability requirements are very strict (< 1%) - to keep the witness phase (and accelerating rate) constant.

Some technical findings

- "Transversal" density modulator for high precision micro-bunching and longitudinal positioning of the driving beam (the use of extremely low transversal emittance for high energy driver).
- By placing proper scattering material inside target slots it is possible to arrange proper longitudinally localized emittance formation.
- The use of pre-microbunching by inversed FEL gives several times growth in driver-to-plasma power transition efficiency. The other attractive option is to use the laser with appropriate micro-structure of the laser pulse. But in any case, the final micro-bunching and creating local emittance matching should be done using a transversal modulator.
- Sequential acceleration with the use of ("straight"!) helical delay-lines for the driver gives the possibility to reach much higher witness energy than that of the driver .

Transversal forces

A very important problem is related to the transversal forces acting on the driver and accelerated beam particles in the excited plasma channel.

These (de)focusing forces are strongly dependent on the phase of plasma oscillations being shifted approximately $\pi/2$ with respect to longitudinal field. And, of course, their effect is the higher, the lower the beam energy is. This problem makes life a lot more difficult and, in particular, pushes to higher driving beam energies.

The maximum of radial force, acting on driver or accelerated beam particles (the sum of eE_{rad} and eB_{az}) is proportional to the electric longitudinal field - accelerating or decelerating. In linear regime of plasma oscillations, this effective transversal field is about 1/3 of the longitudinal field maximum. The rise radius is about $\lambda/2\pi$ (excited plasma channel assumed to be minimal!).

So, the equation for transversal single particle motion will be

$$\frac{d^2}{ds^2}r + \frac{2\pi}{3} \cdot \frac{\text{grad}E_{eV} \, \kappa}{E_{eV} \, \lambda} r = 0 \, ,$$

where s is the longitudinal coordinate, and κ is the fraction of transversal force maximum at the current phase.

If the coefficient is negative (defocusing), the accelerating field about 1 GeV/m leads to incurable defocusing and, hence, to the loss of particles at this phases - no matter how strong, albeit realizable, is the external focusing system.

If this coefficient is positive, effective "plasma beta-function" will be

$$\beta_{pl} = \sqrt{\frac{3}{2\pi} \cdot \frac{E_{eV} \cdot \lambda}{\text{grad}E_{eV} \cdot \kappa}} \, .$$

To direct the driving beam properly, we need appropriate external focusing - by quadruples (the driving beam energy is high!).

Thus, the driving micro-bunch #2 (and even the tail of the first one!) and the following micro-bunches, as well as the witness bunch, travel under combined action of the stationary (for given slice) focusing of the plasma and the alternating focusing of quadruples.

Hence, we have to take care of stability of incoherent transversal oscillations of all useful particles of different energies and at very different plasma focusing - in a time!

The phase structure of accelerating and transversal forces is quite complicated.

At the initial linear stage, their maximums are shifted almost at $\pi/2$. At the developed stage, which is of the main interest, this shift is going down and it is necessary to play with driver and witness bunches phasing and lengths - and even with their angular spreads! - very carefully.

Proper matching of emittances (in size and shape) in every slice of the driver and the witness is necessary.

"Transversal" micro-bunching.

To find the way to introduce a proper micro-bunch structure in the bunches of driving beam (and, also, to prepare a proper witness micro-bunch to be accelerated) is one of the crucial elements of the approach.

The problem looks non-trivial, because we need 0.1 mm range length of each "very high" energy micro-bunch and its positioning should take into account plasma frequency variation (non-linear regime!), hence - micro-bunches should be non-equidistant.

This positioning should prevent parts of driver micro-bunches from both entering the accelerating phases, which would "eat" the plasma oscillations energy instead of pumping it in, and the plasma defocusing phases.

The way we proposed to solve this problem is:
to use extremely low emittance beams and "transversal cutting".

The general layout looks as follows:

At some part of the (future driver) beam channel with high beta value, say, in vertical direction, we arrange local RF structure acting on the traveling bunch with the vertical force linearly depending on the position along the bunch (zero action at the bunch center).

The resulting transversal vertical momentum should be much higher than due-to-emittance internal momenta in the bunch.

When passing the free space long enough, different head-to-tail constituents of each bunch will be positioned differently in vertical direction.

At this distance, a target-cutter is placed, the slots of which are transparent. Other parts of the target destroy beam components completely.

At the same plane where the target-cutter is located a vertically focusing lens is placed (focal length is 2 times smaller than RF section-to-target distance).

At the same distance after the target a similar RF structure is located, which compensates the vertical momenta of bunch components.

Hence, at the exit of this section, each driving beam bunch will be transformed into a series of micro-bunches - properly (longitudinally) shaped and properly positioned!

Placing appropriate thin foils into the slots, we can give each micro-bunch a proper angular spread (and even proper dependence of this spread on slice position along the micro-bunch) - in correspondence to the future local plasma focusing (local beta-value).

Pre-microbunching.

To maximize excitation efficiency of the driving beam, it is worth trying to shift as much particles as possible in the useful phases of each bunch using RF energy modulation - prior the final micro-bunching via sliced target.

The natural option is to use a few wigs IFEL at this wave length. It can easily give energy modulation higher than energy spread in such a high quality beam. The following use of dispersive section will transform energy modulation along the bunch into density modulation of the same wave length.

But let us keep in mind - we need non-equidistant micro-bunching.

Hence, the transversal cutting should follow!

Sequential acceleration.

It is interesting and important to find the way of reaching much higher energies than the driver energy. This energy multiplying is very similar to the usual high voltage klystron driving of linac and to the "two-beam accelerator".
The goal looks this way:

by using sequence of N driving bunches of energy E_{dr}, reach almost proportionally higher energy

$$E_{acc} ==> N \cdot E_{dr}.$$

The way in principle looks quite obvious. One driving (micro-bunched) bunch, via plasma excitation, transfers energy to the accelerated "witness" bunch up to exhaustion; then the next bunch replaces the previous one. And then the process will be repeated N times.

The problem is how to arrange this process in the most efficient and cost effective way.

In my understanding, the best way is to arrange a "helical delay-line" for driving beam train. In this case, we can use a single straight tunnel to host all the beams.

If we use a helix of actual curvature radius R and outer radius r, the delay dL of the train at the helix length L_0 will be

$$\delta L = \frac{L_0 \cdot r}{2 \cdot R}.$$

At the length of driver energy loss, which is about

$$L_0 = \frac{E_{dr}}{gradE},$$

the delay should be equal to the distance between the train bunches dL_b.

The helix curvature is defined by the driver energy E_{dr}, also.

Hence, the requirement for "r" will be

$$r = 2 \cdot \frac{E_{acc}}{B_{hel}} \cdot \delta L_b \quad \text{(CGSE)}$$

where B_{hel} is the helix magnetic field used.

It can be easily seen that this requirement is quite moderate, at least, formally.
<u>All the tolerances are to be analyzed carefully!</u>

When such a helical delay-line is arranged along the whole linear accelerator, the driving beam, bunch by bunch, transfers the energy to the same witness micro-bunch being accelerated.

<u>The influence of particles-plasma scattering.</u>

Not similar to usual accelerators, at PWFA the accelerating channel is filled with plasma particles. Hence, additionally to the coherent fields, beam particles are also influenced by pair collisions. To minimize this influence, it is worth using "100% ionized" hydrogen plasma, which consists of protons and electrons, and we should have in mind this option, only. There are several collision processes which may be of some importance.

- Nuclear interaction of hadrons of driver or accelerated beams.
  The hadron-hadron total cross-section does not exceed

$$\sigma_{hadron} \leq 1 \cdot 10^{-25} cm^{-2}.$$

Hence, the probability to loose particles in 1 km path (in our consideration - per 1TeV) is less than

$$\sigma_{hadron} \cdot n_{pl} \cdot 1km = 10^{-25} \cdot 10^{15} \cdot 10^5 = 10^{-5}$$

So. it is not a limitation in "any" practical conditions.

- The bremstrahlung cross-section even for electrons and positrons is of the same order, so, it is also unimportant. Hence, the single scattering processes are not a problem.
- The most important process - if we do care of emittance growth - is the Coulomb multiple scattering of accelerating particles. The advance of mean square of the particle angle due to this process is

$$\delta\theta_0^2 = \frac{200}{E_{MeV}^2} \cdot \frac{x}{X_0} = 5 \cdot 10^{-24} \cdot \frac{\delta L_{acc} n_{pl}}{E_{MeV}^2},$$

where x/X_0 is the fraction of radiation length passed. Consequently, emittance differential will be

$$\delta\varepsilon_{acc} = 5 \cdot 10^{-24} \cdot \frac{n_{pl}}{gradE} \cdot \frac{\beta_{acc}(E)}{E^2} \cdot \delta(E),$$

where E is the current energy in MeV; and 100% ionized hydrogen plasma is considered. As a result, the current contribution to the final emittance growth will be

$$\delta\varepsilon_{fin} = 5 \cdot 10^{-24} \cdot \frac{n_{pl}}{E_{fin} gradE} \cdot \frac{\beta_{acc}(E)}{E} \cdot \delta E.$$

The final emittance, if influenced by in-plasma collisions only, will make up

$$\varepsilon_{fin} = 5 \cdot 10^{-24} \frac{n_{pl}}{E_{fin} gradE} \cdot \int_{E_{in}}^{E_{fin}} \frac{\beta_{acc}(E)}{E} \cdot dE.$$

This result shows the most important limitation for electron-positron collider based on ultimate gradient PWFA.

Plasma arrangements

Plasma along the whole pass of the beams should be prepared in advance. As was mentioned above, to minimize the plasma binary collisions influence, especially on accelerating particles for collider use, it is preferable for the plasma to be the hydrogen one, and the ionization degree should be close to 100%. Let us evaluate the energy needed for such arrangement:

$$E_{source} = \pi R_{pl}^2 n_{pl} L_{acc} \eta^{-1} E_{ion},$$

where

E_{ion} - ionization potential,

η - efficiency of energy transfer from ionization source,

R_{pl} - radius of plasma channel.

Example:

$$E_{source} = \pi \cdot (3mm)^2 \cdot 10^{15} \cdot 10^5 \cdot 10^2 \cdot 10eV \cdot 2 \cdot 10^{-19} = 6 \frac{kJ}{TeV}$$

At repetition rate "f", the power for ionization will be proportional to "f".

But the most difficult plasma related requirement seems to be the necessity to keep its density constant along the whole accelerator - in 1% range (to keep full accelerating gradient at the chosen witness phase).

PWFA based electron-positron collider.

Let us discuss in a very preliminary way the option of linear collider based on PWFA approach.

The general layout which I had in mind looks as follows:
- 11 GeV electron "driver" accelerator produces trains of bunches.
- Each train contains 100 bunches; each 1 cm long, 10^{11} particles in each bunch, bunches are separated by one or several accelerator wave length.
- Each bunch, by using the pre-buncher and the transversal cutting, is transformed into 10 microbunches of the length 0.2 mm each.
- The last bunch of the train (#100) is directed into the plasma of appropriate density; other bunches enter a spiral isochronous delay-line. Traveling through the plasma with properly arranged quadrupole lenses structure inside (0.5 cm long lenses per each 10 cm), the fine-cut bunch excites the accelerating field of 1 GeV per meter.
- With some short delay (half the plasma wave length to accelerate electrons, full plasma wave length to accelerate positrons), the driving bunch is followed by the "witness" (the accelerated bunch) of the length 0.03 mm, of the "same" energy, and containing up to $2 \cdot 10^9$ particles.
- After the last microbunch of the driving bunch #100 passed 10 m section and got decelerated to 1 GeV, the bunch is carefully replaced by the previous one, i.e., by #99; the micro-bunch under acceleration, of energy +10 GeV, is positioned the same way to the new bunch.

This cycle is repeated for 100 times, bringing the witness to 1 TeV - upon 1 km of acceleration. These super-cycles are repeated at 10 kHz frequency. The same arrangements and processing are made for the counter beam. So, we get a 1 TeV + 1 TeV collider (electron-positron, photon-photon, electron-proton).

Using formulae on plasma scattering and plasma focusing for more or less optimized conditions, we can show, that geometrical emittance at 1 TeV of 10^{-13} cm·rad is reachable (one should bear in mind rigorous tolerances!).

So, there is a hope to get 10^{34} cm^{-2}sec^{-1} with driving beams of several tens of MW mean power, at least, for photon-photon and photon electron collisions.

Proton klystron.

With the use of intense and very low emittance beams of super-high energy proton storage rings / colliders, it looks feasible to drive PWFA linacs for reaching energy multiplication for ejected protons (at least, for neutrino experiments);

to accelerate pre-cooled antiprotons; electrons, positrons; ionizationally pre-cooled muons; pions; and even kaons (relativistic kaons live long enough to be accelerated by energy gradients close to 1 GeV/m).

Novosibirsk PWFA experiment under preparation.

The test experiment is under preparation at INP to study PWFA, with hope to reach more than 0.5 GeV/m at the several tens of cm (see the figure).

Layout of VEPP2M-PWFA complex

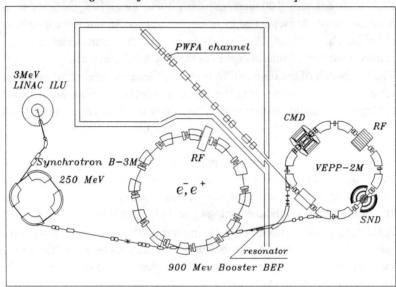

The Open Waveguide Structure

R.H. Pantell

310 McCullough Building, Stanford University, Stanford, CA 94305

Abstract. An open waveguide provides a useful structure for laser accelerators. It shall be shown that a simple structure, with dimensions that are 3–4 orders of magnitude larger than the laser wavelength, supports a mode with a strong on-axis longitudinal field, with low loss, and with small phase slippage relative to an electron moving at close to the velocity of light.

STRUCTURE DESIGN

The open waveguide [1,2], illustrated in Figure 1, provides a useful structure for laser acceleration. It consists of a series of apertures of diameter $2a$ with cylindrical symmetry, separated by a distance b. The modes for this structure are identical to the modes of a Fabry Perot interferometer with planar mirrors of diameter $2a$ and a length b apart. In the original modal analysis [3] the longitudinal field component was not of interest and not evaluated. It shall be shown that this simple structure, with dimensions that are 3–4 orders of magnitude larger than the laser wavelength, supports a mode with a strong on-axis longitudinal field, with low loss, and with small phase slippage relative to an electron moving at close to the velocity of light.

ANALYSIS

The modes of the open waveguide may be obtained from the Kirchoff diffraction integral [4] wherein the transverse field of a linearly polarized mode at the $(i+1)$ aperture E_{i+1} may be expressed in terms of the transverse field E_i at the ith aperture [3]

$$E_{i+1}(\rho_2) = -M \int_0^1 \rho_1 d\rho_1 E_i(\rho_1) J_1(M\rho_1\rho_2) \operatorname{Exp}\left[-jM\frac{\rho_1^2 + \rho_2^2}{2}\right] \quad (1)$$

where $\rho_i = \frac{r_i}{a}$, $M = 2\pi N$, $N = $ Fresnel number $= \frac{a^2}{\lambda b}$, and J_1 is the Bessel function of the first kind and first order. Equation (1) is valid for

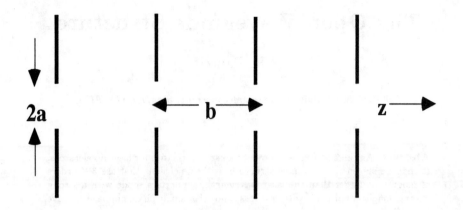

FIGURE 1. An open waveguide structure for laser acceleration. It consists of a series of apertures, where the function of each aperture is to remove a portion of the diffracted wave from one section to the next. This structure is equivalent to a Fabry Perot interferometer with circular, planar mirrors having diameter $2a$ and separated by a distance b.

$$\frac{b}{a} \gg 1 \quad \text{and} \quad \left[\frac{b}{a}\right]^2 \geq 5N \qquad (2)$$

which are conditions that have been applied to approximate the distance between a point on one aperture and a point on the adjacent aperture. The geometric phase shift factor $\exp(-jkb)$ has not been included on the right-hand side of Eq. (1).

Equation (1) has been solved both by an iteration procedure, calculating successive E_i, and also as an eigenvalue problem by writing Eq. (1) in matrix form. The former approach gives the transient evolution of the field from initial excitation to steady-state, and the latter gives a direct evaluation of the complex eigenvalues.

To obtain a waveguide mode suitable for particle acceleration, the guide is excited with the lowest order Hermite-Gaussian mode with an on-axis, longitudinal field component:

$$E_z(x,y) = E_0 \frac{x}{w} \exp\left[-\frac{(x^2 + y^2)^2}{w^2}\right] \qquad (3)$$

Figure 2(a) illustrates the transverse field amplitudes for the initial and steady-state fields as functions of the normalized distance x/a, where $w = 0.7a$ and $N = 48$. The fields have been normalized to unity at their maximum values. In Figure 2(b) the abscissa scale has been expanded to show the increase in

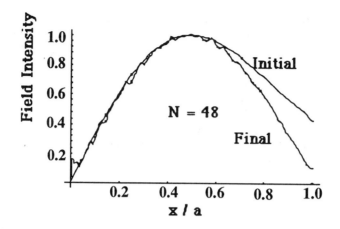

Figure 2(a)

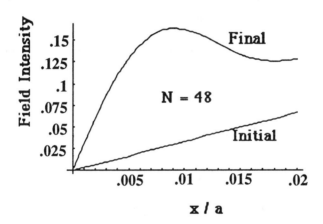

Figure 2(b)

FIGURE 2. Figure 2(a) shows the initial and final transverse field intensities, normalized to unity at their respective maxima, as functions of x/a for a Fresnel number of 48 over the entire aperture. Figure 2(b) uses an expanded abscissa scale showing the final field near $x = 0$.

the field gradient near the origin, which translates into an increase in the longitudinal field component. From Figure 2(a) it is seen that the amplitude at $x = a$ decreases significantly, with a consequent reduction in the likelihood of material damage to the apertures.

From $\nabla \cdot E = 0$, the longitudinal field E_z may be determined, and Figure 3 shows $ka|E_z|$ as a function of x/a. The value for E_z has been calculated for an E_x that is normalized to unity at its maximum value, but the actual value of E_z depends upon the mode power P. In addition, for a given P, $|E_z|$ can be increased by $\sqrt{2}$ by adding a y-polarized mode to the x-polarized mode to obtain a radially polarized mode. For the latter case we find that

$$\frac{a^2}{\lambda} = 1.6 \times 10^2 \frac{P^{.5}}{|E_z|} \qquad (4)$$

from which the aperture dimension may be calculated with specified values of P, E_z, and λ. The spacing between apertures b is given by $b = \frac{a^2}{N\lambda}$. Choosing $P = 25$ TW and $|E_z| = 1.5$ GV/m, for $N = 48$ and $\lambda = 1.0 \mu$m we have that $2a = 1.4$ mm and $b = 1.2$ cm. The inequalities specified by Eq. (2) are satisfied for these parameter values. The fact that the structure dimensions are 3–4 orders of magnitude larger than the wavelength simplifies fabrication.

Modifying the structure by replacing the planar apertures with truncated cones, as illustrated in Figure 4, does not change the modes if the cone angle Θ is $\gg \Theta_d \approx 0.6\lambda/a =$ aperture diffraction angle. The advantage to using cones

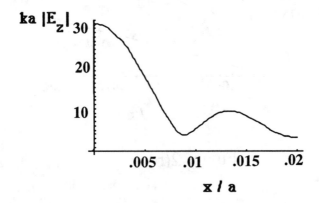

FIGURE 3. Magnitude of the longitudinal field component for the final field as a function of x/a. The longitudinal field was obtained from $\nabla \cdot E = 0$ using the normalized transverse field shown in Figure 2.

is that for a specified power density in a wave propagating in the z-direction the fluence on the cone surfaces is reduced, thereby increasing the laser power threshold for cone damage. With the previously specified parameters and $\Theta = 10$ mrad, the maximum fluence on the first aperture is $0.55\ J/cm^2$ and this falls to $0.1\ J/cm^2$ after the second aperture. Damage threshold is about $2\ J/cm^2$ [5]. Since the function of the cones is solely to remove the portion of the wave at radii greater than the aperture, cone damage is important only if holes are drilled through the material.

From the eigenvalue solution an approximate expression may be obtained for the wave propagation in the open guide of the form $\exp[i(\omega t - kz + \phi) + \xi z]$ where ϕ is the initial phase and ξ is given by the complex eigenvalue. If the electron energy is ≥ 10 GeV, then $(\omega t - kz)$ is ≈ 0 over the length of interest and

$$\xi = i\frac{\lambda}{a^2} - \frac{0.24}{N^{.4}}\frac{\lambda}{a^2} \qquad (5)$$

From Eq. (5) the energy transfer W from the wave to the electron beam may be determined as a function of z, and the result is shown in Figure 5(a) where $\frac{W(eV)}{\sqrt{P(W)}}$ is plotted vs. $\frac{\lambda z}{a^2}$. In moving from Point A to Point B there is 180° phase slippage between the particle and the wave. At Point B the electron is in a position where the field is zero and beyond this point the electron loses energy. Energy transfer oscillates with decreasing amplitude due to power loss on the structure.

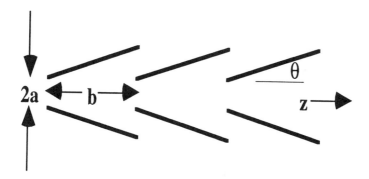

FIGURE 4. The planar apertures drawn in Figure 1 are replaced with truncated cones. The waveguide modes remain unaltered, whereas the damage threshold increases.

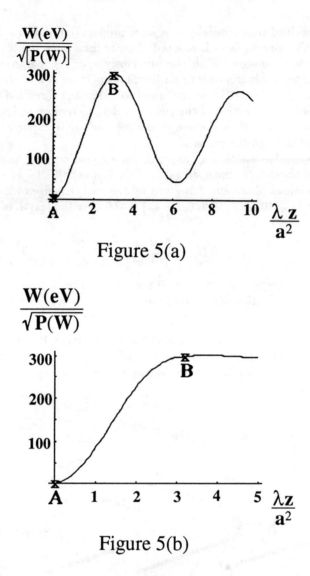

FIGURE 5. Energy transfer as a function of interaction length. Figure 5(a) is energy transfer along a structure of undetermined length, with 180° phase slippage from Point A to Point B. Figure 5(b) is energy transfer along a structure terminated at Point B (roughly 150, 1.2 cm sections). The optical field is propagated in free space following termination of the structure, and its effects on the electron beam are included.

If the structure is terminated at Point B the wave then diffracts into free space and there is little additional energy transfer, as shown in Figure 5(b). For $P = 25TW$, the energy transfer is 1.5 GeV in a distance of 1.5 m.

The analysis used to obtain Figure 5 assumed the steady-state solution, but over the first 1.5 m there are significant field amplitude fluctuations due to the presence of higher order modes which are quite large even though they attenuate more rapidly than the lowest-order mode. Incorporating the transient amplitude and phase variations it is found that the energy transfer is 1.6 GeV in 1.6 m, a result that is not very different form the steady-state solution.

SUMMARY

The open waveguide structure has a number of interesting features as a laser accelerator:

(a) The structure is simple with reasonable dimensions.

(b) Interaction is in vacuum.

(c) The gradient is high.

(d) Electron motion is linear.

(e) There is high energy transfer per stage. For $P = 25TW$, $\Delta W = 1.6$ GeV in 1.6 m.

Problems for future consideration are:

(a) The manner in which individual stages would be combined. It is desirable to minimize the distance between stages to maintain as high a spatially-averaged gradient as possible.

(b) Excitation of the open guide with a Fabry Perot interferometer using planar mirrors and the same N as the guide. Then, the form of the excitation would be identical to the steady-state waveguide mode so that there would not be a transient region.

(c) The effect of finite electron beam radius on emittance and energy spread.

(d) Mode modification due to the finite thickness of the cones.

ACKNOWLEDGMENT

The author is grateful to John Lewellan and Dr. Joseph Feinstein for their important contributions.

REFERENCES

1. L.A. Weinstein, *Open Resonators and Open Waveguides*, (The Golem Press, Boulder, Colorado, 1968).
2. G. Goubau, in *Advances in Microwaves Volume 3*, 67, edited by L. Young, (Academic Press, NY, 1968).
3. A.G. Fox, T. Li, Bell System Technical Journal **40**, 453 (1961).
4. M. Born and E. Wolf, *Principles of Optics*, 4th edition, 370 ff (Pergamon Press, London, 1970).
5. B.C. Stuart, M.D. Feit, A.M. Rubenchik, B.W. Shore, M.D. Perry, Phys. Rev. B **53** 1749 (1996).

Self-Modulation of High-Intensity Laser Pulses in Underdense Plasmas and Plasma Channels

N.E. Andreev*, L.M. Gorbunov**, V.I. Kirsanov*, and A.S. Sakharov***

*High-Energy Density Research Center, Joint Institute for High Temperatures, Russian Academy of Sciences, Izhorskaya 13/19, Moscow 127412, Russia

**Lebedev Physical Institute, Russian Academy of Sciences, Leninskii pr. 53, Moscow 117924, Russia

***General Physics Institute, Russian Academy of Sciences, Vavilova st. 38, Moscow 117942, Russia

Abstract. The analysis is carried out for the basic regimes of the self-modulational instability of high-intensity ($I \sim 10^{17}$-10^{18} W/cm^2) laser pulses in underdense ($\omega_0 >> \omega_p$) plasmas. The conditions under which these basic regimes dominate, growth rates corresponding to these regimes, and phase velocity of the plasma wave excited due to the instability are discussed in relation to the previous and possible future experiments on laser acceleration of electrons in the configuration utilizing the self-modulation of laser pulses.

I. INTRODUCTION

Laser accelerators [1, 2] are considered as one of the most promising approach to reach extremely high accelerating gradients, and the recent experiments have confirmed the possibility to obtain acceleration of electrons with an accelerating rate as high as 100 GeV/m [3-5]. Because, in these experiments, the region of acceleration was less than 1 mm, the final energy of the accelerated particles did not exceed 100 MeV. The primary goal of the future experiments is to demonstrate the acceleration of electrons up to energies exceeding 1 GeV. The main obstacle on this way is the limitation arising due to the transverse evolution of the pulse and, primarily, the diffraction of the laser pulses (see for example [2, 6]); as a result, the actual distance over which the electrons can be accelerated is about $2Z_R$ (where Z_R

is the Rayleigh length). To avoid this limitation, the special configurations of the laser accelerators were proposed [7-13] among those the most promising are the configurations that assume the use of plasma channels providing a long-range transportation of the laser pulse (over the distances substantially exceeding Z_R) [10-13].

The other opportunity to increase the energy of accelerated electrons is to increase the laser sport size (decreasing, consequently, the laser-field intensity and increasing the length at which the pulse diffracts); in this case, high values of the excited wake field can be obtained at the expense of an increase in the pulse energy [14] if the pulse modulation (and, consequently, the resonance excitation of the plasma wave) due to the self-modulation instability is provided [2, 15–24]. Note that, this increase in the sport size is also accompanied by the increase in the transverse dimension of the wake field and decrease in the radial component of the wake field; this is advantageous for both increasing the number of electrons that can be loaded and accelerated in the wake field, and decreasing the transverse emittance because of the decrease in the ratio between the transverse and longitudinal components of the wake field.

The instability of short powerful laser pulses propagating in an underdense plasma ($\omega_0 >> \omega_p$) is known to cause the axial modulation of the laser-pulse intensity with the characteristic (resonance) wave number $k_p \approx \omega_p / c$ (see, [2, 8-11, 14-24]). Depending on the laser pulse (intensity and geometrical dimensions) and plasma (density) parameters, this self-modulation can be either one- or three-dimensional. Though the one- and three-dimensional mechanisms differ, their common feature is the feedback loop involving the excited wave of the electron plasma density with the wave number close to the resonant one (k_p) and the wave of the longitudinal laser-intensity modulation. The important consequence of the instability onset is the generation of the strong wake-field behind the pulse, which can be used for electron acceleration [1].

The seed for this instability can arise either due to the evolution of the laser pulse [8-10, 20] or can be provided by sharp leading edge of the pulse [14, 19, 22, 23] or an additional low intensity pulse with shifted frequency (beating of the field of this pulse and the main one is a seed for the instability) [11].

In spite of a large variety in the regimes of self-modulation [2, 15-17, 21, 22], in the real experimental situations only some of them are most probable to occur. Consequently, here, we separate the basic regimes that can occur under the conditions of reported or planned experiments on self-modulation.

I. BASIC EQUATIONS

To describe the self-modulation of a laser pulse and accompanying excitation of the plasma wave we use the set of equations for the normalized envelope of the laser field $a = eE_0/m_e c\omega_0$ ($E = (1/2)[E_0 \exp(-i\omega_0 t + ik_0 z) + c.c.]$) and perturbations of electron plasma density $N = (n - n_0)/n_0$ in the form [16]

$$\left(2i\omega_0 \frac{\partial}{\partial t} + c^2 \Delta_\perp + c^2 \frac{\omega_p^2}{\omega_0^2} \frac{\partial^2}{\partial \xi^2} + 2c \frac{\partial^2}{\partial \xi \partial t}\right) a = \omega_p^2 \left(N + \frac{r^2}{R^2} - \frac{1}{4}|a|^2\right) a, \quad (1)$$

$$\left(\frac{\partial^2}{\partial \xi^2} - \frac{2}{c}\frac{\partial^2}{\partial \xi \partial t} + k_p^2\right) N = \frac{1}{4}\Delta |a|^2. \quad (2)$$

The system of equations (1) and (2) is obtained for $|a|, |N| \ll 1$ and $\partial \ln a/\partial t, \partial \ln N/\partial t \ll \omega_p$. Here, the coordinates $\xi = z - v_g t$ and t correspond to the frame of reference moving with the pulse whose group velocity in the linear approximation is $v_g = c^2 k_0/\omega_0$ (the corresponding γ-factor is equal to $\gamma_g = \omega_0/\omega_p$). The term r^2/R^2 incorporates a possible transverse plasma inhomogeneity in the form of a parabolic plasma channel ($n = n_0(1 + r^2/R^2)$) with the characteristic radius $R \gg k_p^{-1}$ (the limiting case $R \to \infty$ corresponds to a homogeneous plasma).

In order to analyze the conditions under which different regimes of self-modulation of a laser field can occur, we treat the problem by using a simplified approach allowing us to consider both one- and non-one-dimensional regimes by solving the same equations. We consider the pulse localized in the region $-L < \xi < 0$ and assume that its envelope is initially uniform $a_0 = const$. We introduce two-dimensional perturbations of the electron plasma density in the form

$$N = \frac{1}{2}\left(\hat{N}(\xi, t) \exp(ik_p \xi + ik_\perp y) + c.c.\right). \quad (3)$$

We also introduce the perturbations of the pulse envelope $\delta a = a - a_0$ in the similar form. The characteristic transverse wave number is denoted here as $k_\perp$. For perturbations of the chosen form, from (1) and (2), the equation for the evolution of the density perturbations follows:

$$\left(\frac{\partial}{\partial \xi} - \frac{1}{c}\frac{\partial}{\partial t}\right)\left[\frac{\partial^2}{\partial t^2} + \frac{c^4}{4\omega_0^2}\left(k_\perp^2 + \frac{\omega_p^2}{\omega_0^2}k_p^2\right)^2\right]\hat{N}(\xi,t) =$$

$$= -ia_0^2 \frac{\omega_p^3}{16\omega_0^2} ck_\perp^2 \hat{N}(\xi,t) - ia_0^2 \frac{\omega_p^3}{16\omega_0^2}\left(\frac{\omega_p^2}{\omega_0^2}ck_p^2 - 2ik_p\frac{\partial}{\partial t}\right)\hat{N}(\xi,t). \quad (4)$$

Here, we need to clarify the relation between the used simplified approach and the case of a real pulse whose field is spatially localized (with the typical scale $L_\perp$) in the transverse direction. To obtain qualitative results in the simplified approach, we associate the typical scale of the transverse inhomogeneity of the pulse intensity with a typical scale of the harmonic perturbations assuming these scales to be close (i. e., $L_\perp \sim k_\perp^{-1}$), which corresponds to considering a class of so-called "whole-beam" instabilities [2]. This approach is instructive for clarifying the basic features of the self-modulation instability.

For such a class of instabilities, the quantitative relation between $L_\perp$ and $k_\perp^{-1}$ can be obtained by comparing equation (4) with the equation describing self-modulation in the paraxial approximation (see, for example [19, 21]). For the axially symmetric laser pulse with a Gaussian transverse profile $a = a_0 \exp(-r^2/L_\perp^2)$, this equation has the form:

$$\frac{\partial}{\partial \xi}\left(\frac{\partial^2}{\partial t^2} + v^2\right)\hat{N} = -ia_0^2 \frac{\omega_p^3}{\omega_0^2}\frac{c}{L_\perp^2}\hat{N}, \quad (5)$$

where

$$v^2 = \left(3 + \frac{a_0^2 k_p^2 L_\perp^2}{8} + \frac{k_p^2 L_\perp^4}{4R^2}\right)\frac{4c^4}{\omega_0^2 L_\perp^4}. \quad (6)$$

In contrast to equation (4), in deriving (5), we adopted the quasistatic approximation for the excited density perturbations ($\partial \hat{N}/\partial t \ll c\partial \hat{N}/\partial \xi$), i.e., assumed that the solution weakly changes while the pulse passes the distance equal to the pulse length. In this approximation (i.e., when the time derivative in the first operator on the left-hand side is small) and, additionally, when the time derivative on the right-hand side of Eq. (4) is small, the equations (4) and (5) have identical form.

Comparing in equations (4) and (5) the main terms related to transverse effects [first term on the right-hand side of (4) and the right-hand side of (5)], we find that the quantitatively correct expression for the three-dimensional self-modulation can be obtained from (4) if we assume $k_\perp = 4L_\perp^{-1}$.

III. BASIC REGIMES OF THE INSTABILITY

For a regular acceleration of electrons to high energies, the relativistic factor corresponding to the phase velocity of the excited plasma wave $\gamma_{ph} = (1 - v_{ph}^2 / c^2)^{-1/2}$ is to be high. For a laser-driven wake field, the phase velocity is close or less than the group velocity of the laser pulse v_g [2, 21]. To reach GeV energy range of accelerated electrons, the relativistic factor corresponding the phase velocity of the wake wave should exceed 30 ($\gamma_{ph}^2 > 10^3$) and the relativistic factor corresponding to the group velocity of the driving pulse, $\gamma_g = (1 - v_g^2 / c^2)^{-1/2} = \omega_0 / \omega_p$, is to be the same or even higher.

A further simplification of the problem is related to the fact that, for a sufficiently high γ_g, the inequality

$$L_\perp k_p < 4 \frac{\omega_0}{\omega_p} \equiv 4\gamma_g \qquad (7)$$

is satisfied and, correspondingly, we can neglect on both sides of (4) the terms proportional to $(\omega_0 / \omega_p)^2 k_p^2$ as small compared to those containing $k_\perp^2$. Physically, this corresponds to neglecting the dispersion effects compared to the diffraction effects. This is typical of the parameters of the present and possible future experiments on laser–plasma electron acceleration using the configuration with laser-pulse self-modulation.

The asymptotic solution of equation (4) (which can be obtained by using the Laplace method and evaluating the inversion integral by the saddle-point method) allows us to determine both the amplitude $|N|$ and phase φ of the excited plasma density wave, separate typical regimes of the instability, and determine the conditions of their occurrence. In the frame of reference related to the pulse (corresponding to the ξ and t variables), the temporal variation of the phase determines the deviation of the phase velocity of the wave from the group velocity of the pulse ($v_{ph} - v_g = -k_p^{-1} \partial\varphi(\xi,t) / \partial t$) and, therefore, the γ-factor corresponding to the phase velocity of the plasma wave [$\gamma_{ph} = \gamma_g (1 + 2\gamma_g^2 \omega_p^{-1} \partial\varphi / \partial t)^{-1/2}$].

The obtained solution shows that there are several basic regimes (stages) that can occur during the onset of the instability. Figure 1 shows the regions corresponding to these regimes in the plane of variables ct / Z_R and $k_p |\xi| (P / P_c)$, where $Z_R = k_0 L_\perp^2 / 2$, P is the pulse power, and $P_c [\text{GW}] = 17 \gamma_g^2$ is the critical power for the relativistic self-focusing ($P / P_c \approx a_0^2 k_p^2 L_\perp^2 / 32$).

The region above the horizontal dashed line corresponds to the times greater than the time during which the pulse passes the Rayleigth length, and the solutions of the equation (4) for this region have a sense only if the some guiding structures are used for the pulse transportation (see [21]).

In the region below the dotted line, the quasistatic approximation [7] is not valid and the solution describing the self-modulation instability in this region needs a treatment with allowance for the time derivative in Eq. (2) [and, consequently, in the operator $\partial/\partial\xi - c^{-1}\partial/\partial t$ in (4)] [17, 18].

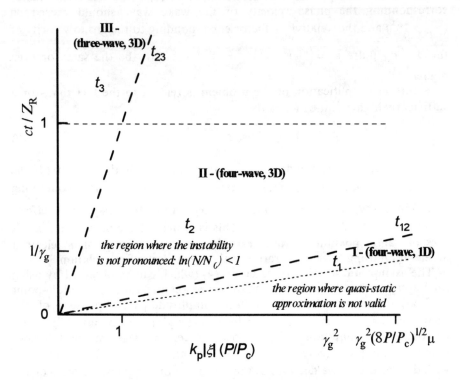

FIGURE 1. Schematic diagram for the regions corresponding to the basic regimes of the self-modulation instability in the coordinate–time plane. The characteristics of the regimes and the boundaries of the regions in which the regimes occur are presented in Table. Notation $\mu = L_\perp k_p / 4\gamma_g$ is used.

The region between the abscissa axis ($t = 0$) and solid line corresponds to small amplification factors of the initial perturbations ($K = |N/N_0| \sim 1$) and is of minor interest.

Two bold dashed lines separate three basic regimes of the instability. The intersections of these lines with the solid line determines two basic reference points [(1,1) and $(\gamma_g^2, \gamma_g^{-1})$] in the diagram. To the left of the additional reference point on the horizontal axis, $\gamma_g^2 (8P/P_c)^{1/2} (k_p L_\perp / 4\gamma_g)$, the quasi-static approximation is always valid for the description of the instability (for $\ln K > 1$).

Note that the diagram in Fig. 1 allows us to identify the regime of the instability in every position of the pulse (at different positions, the different regimes can be dominating). The field behind the pulse is of most interest for the acceleration of electrons, and the regime of the instability dominating in the rear part of the pulse (near the trailing edge of the pulse, i. e., at $|\xi| = L$) determines this wake field. The amplitude and phase velocity of the wake wave equal to their values at the trailing edge of the pulse.

Table summarizes the characteristics (including the γ-factor corresponding to the phase velocity of the excited plasma wave) of the instability for the basic regimes and the conditions for their realization. For brevity, the notation $\eta = k_p |\xi| (P/P_c)$ is used.

TABLE. Basic parameters of the regimes

Regimes	I (four-wave, quasi-1D)	II (four-wave, 3D)	III (three-wave, 3D)
Time domain of the regime occurrence	$t > t_1 = \gamma_g Z_R / c\eta$ $|\xi|/c < t < t_{12} =$ $= Z_R \eta / c\gamma_g^3$	$t > t_2 = Z_R / c\eta^{1/2}$ $Z_R \eta / c\gamma_g^3 = t_{12} < t < t_{23} =$ $= Z_R \eta / c$	$t > t_3 = Z_R / c\eta$ $Z_R \eta / c = t_{23} < t$
$\ln K$	$2(2\eta c t / Z_R \gamma_g)^{1/2}$	$(3^{3/2}/2)(2\eta c^2 t^2 / Z_R^2)^{1/3}$	$2(2\eta c t / Z_R)^{1/2}$
$\gamma_g^2 / \gamma_{ph}^2$	$1 + \dfrac{16\gamma_g^2}{k_p^2 L_\perp^2}$	$1 + \dfrac{16\gamma_g^2}{k_p^2 L_\perp^2} \dfrac{1}{4\gamma_g} \left(\dfrac{2\eta Z_R}{ct} \right)^{1/3}$	$1 + \dfrac{16\gamma_g^2}{k_p^2 L_\perp^2} \dfrac{1}{2\gamma_g}$

For the regime I, the term with a time derivative on the right-hand side of (4) is dominating [in the initial equations, this term corresponds to the term with a cross-derivative on the left-hand side of equation (1)]. The description of this regime with neglecting transverse and dispersion effects was carried out in [17, 18] where the expression for time-space growth of the perturbations corresponding to this regime was obtained from which it follows that the phase velocity of the plasma wave is equal to the group velocity of the pulse. Incorporation of three dimensional effects does not noticeably change the amplitude characteristics of the instability, but results in the large change (under the conditions when inequality (7) is satisfied) in the γ_{ph}, which turns out to be less than γ_g. Note, that because the laser-pulse modulation corresponding to this regime is still related to the longitudinal redistribution of the pulse energy, this regime can be treated as a quasi-one-dimensional. In the quasistatic approximation, this regime can occur only if the condition $8(P/P_c) > 16\gamma_g^2 / k_p^2 L_\perp^2$ is satisfied. Also, it worth noting that the incorporation of the diffraction term [first term in the parenthesis on the right-hand part of Eq. (4)] can also lead to an additional reduction in γ_{ph}; however, this reduction is negligibly small for the case we consider here [inequality (7)].

For $\eta < \gamma_g^2$ (i.e., for $32(\omega_0/\omega_p)^2/(k_p L_\perp)^2 > a_0^2 k_p L$, see [21, 22]), the instability is essentially three dimensional. A strong amplification of initial perturbations during the time less than Z_R/c is possible only for $\eta > 1$. This corresponds to the regime II of the instability. In this regime, in contrast to the regime I (as well as to the regime III), the asymptotic solution (for $\ln K > 1$) shows that the value of the phase velocity of the excited plasma wave varies (increases) during the onset of the instability and remains below v_g (see Table).

Since the regime III requires the times greater than Z_R/c for a substantial growth of the initial perturbations, it corresponds to the transportation of a pulse in a guiding structure. The expressions presented for this regime in Table, assumes that the matched parabolic plasma channel ($v = 2c/Z_R$ in (5), see [19, 20]) is used for the pulse transportation. In this regime, γ_{ph} is constant and can be much smaller than γ_g (see Table).

We mention that there is a virtual possibility of the realization of one more regime (so-called short pulse regime, $ct/Z_R \gg (P/P_c)^{-1}(k_p\xi)$ [2]), which can occur even later than III regime at times much greater than Z_R/c. However, the realization of this regime is possible only for the extremely small initial perturbations and it requires the presence of a plasma channel.

Under the conditions of the present experiments involving the self-modulation [3-5] ($\omega_0/\omega_p \sim 10$), the parameter $\eta = k_p|\xi|(P/P_c)$ lies near or to the right of the value γ_g^2. In this case, according to our analysis, the instability onset starts from the 1D regime (regime I) and can pass to the 3D regime (regime II). However, for less dense plasmas ($\gamma_g^2 > \eta$), our diagrams predicts that the mechanism for the instability will be three dimensional from the beginning.

Therefore, because the substantial increase in the energy of accelerated particles requires the use of less dense plasmas than those in the existing experiments, i.e., $\gamma_g \gg 10$, it is the case of the 3D instability (regime II) that is of interest for future accelerating experiments in homogeneous plasmas. In the next Section, we will study this regime in more detail theoretically (in the linear paraxial approximation) and numerically.

IV. SELF-CONSISTENT MODELING OF WAKE FIELD GENERATION

In this section, we consider analytically and numerically the self-modulation of a flat-top pulse whose leading edge is sharp and excites a plasma wave that is a seed for the instability. The analytical solution [22, 23] corresponding to the regime II of the equation (5) is obtained in the paraxial approximation. We compare this solution and the results of numerical calculations [23] based on the set of equations (1) and (2). The initial radial profile of the pulse is assumed to be Gaussian, $a = a_0 \exp(-r^2 / L_\perp^2)$, and the pulse and plasma parameters are defined by the following relations $\gamma_g^2 = 450$, $P/P_c = 0.68$, $k_p L = 54$, $L_\perp = 4.5\lambda_p$, and the size of the pulse leading edge is less than k_p^{-1}.

The wave of the plasma density perturbations excited by the leading edge of the pulse, which is a seed for the instability is clearly seen in the Fig. 2b. The profile of the laser pulse (Fig. 2a) is not modulated at the initial moment. Figs. 2c and 2d show the profile of the same pulse and excited perturbations of the plasma density after the pulse has passed the distance equal to $0.6Z_R$; at this moment, the regular modulation of the pulse is strongly pronounced and the amplitude of the excited wake field is close to its maximum value.

Figure 3 demonstrates that, in the nonlinear stage when the strong wake field is excited, the γ-factor corresponding to the phase velocity of the excited plasma wave is close to γ_g. It is seen from the figure that, in the linear stage of the instability, there is a good correspondence between the numerical solution of equations (1) and (2) and analytical solution obtained in the paraxial approximation. We note that, in the linear stage of the instability, the initial decrease in the phase velocity of the excited plasma wave (for $\ln K \ll 1$) is followed by the subsequent increase (see Fig. 3). This can be advantageous for trapping the accelerated electrons into the plasma wave.

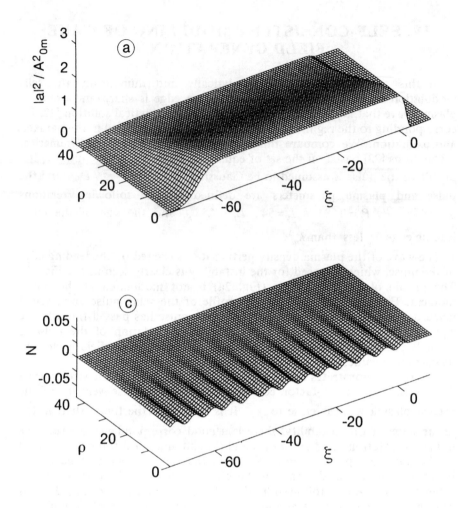

FIGURE 2 (Plots a and c). The spatial (radial and axial) distributions of the pulse amplitude and density perturbation at the initial moment.

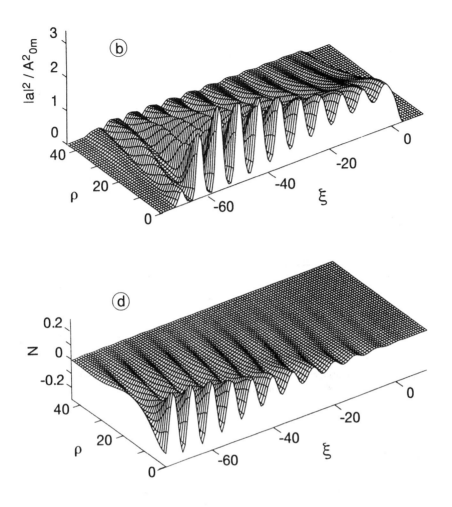

FIGURE 2 (Plots b and d). The spatial (radial and axial) distributions of the pulse amplitude and density perturbation at $t = 0.6 Z_R/c$.

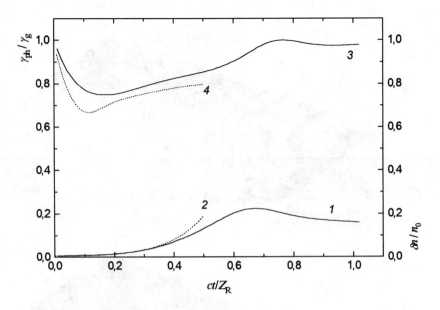

FIGURE 3. Temporal variations of the amplitude of the density perturbations $\delta n / n_0$ (curves *1* and *2*) in the wake field and γ_{ph} / γ_g (curves *3* and *4*). Solid curves *1* and *3* shows the numerically obtained results and the dotted curves *2* and *4* present the results of the linear theory (paraxial approximation).

V. Conclusion

We have shown that, for realistic conditions of accelerating experiments, a large variety of the regimes of self-modulation of the laser pulse is limited to only three basic regimes and, for the future experiments on electron acceleration (in the configuration with the laser-pulse self-modulation in an homogeneous plasma), only a regime that is essentially three dimensional (II-regime: 3D, four-wave) is of interest as providing the further increase in the energy of accelerated electrons. We studied in detail this regime and found that, in the initial stage of the instability, the γ-factor corresponding to the phase velocity of the plasma wave is smaller than that corresponding to the group velocity of the laser pulse. However, fortunately for the acceleration, in the stage of the developed instability, these γ-factors are fairly close.

We considered the equation that allows us to study one and three-dimensional effects simultaneously. This makes it possible to incorporate the three-dimensional effects in the consideration of the regime that was previously always considered with neglecting these effects (regime I). We have shown that, when $4\gamma_g > k_p L_\perp$, under the conditions corresponding to this quasi-one-dimensional regime, γ_{ph} of the excited plasma wave is less than γ_g.

ACKNOWLEDGMENTS

One of the authors (N.E.A.) acknowledges many insightful and stimulating discussions with E. Esarey, C. Joshi, T. Katsouleas, W.B. Mori, and Z. Parsa.

REFERENCES

1. Tajima, T. and Dawson, J.M., *Phys. Rev. Lett.*, **43**, 267–270 (1979).
2. E. Esarey, P. Sprangle, J. Krall, and A. Ting, *IEEE Trans. on Plasma Sci.* **24**, 252–288 (1996).
3. Nakajima, K., Fisher, D., Kawakubo, T., Nakanishi, H., Ogata, A., Kato, Y., Kitagawa, Y., Kodama, R., Mita, K.,Shiraga H., Suzuki K., Yamakawa, K., Zhang, T., Sakawa, Y.,*Phys. Rev. Lett.* **74**, 4428–4431 (1995).
4. Modena, A., Najmudin, Z., Dangor, A.E., *et al.*, *Nature* **377**, 606–608 (1995); *IEEE Trans. on Plasma Sci.* **24**, 289-295 (1996).
5. Umstadter, D., Chen, S.-Y., Maksimchuk, A., et al., *Science* **273**, 472-475 (1996).
6. W. Leemans, C.W. Siders, E. Esarey, N. Andreev, G. Shvets, and W.B. Mori, *IEEE Trans. on Plasma Sci.* **24**, 331–342 (1996).
7. Sprangle, P., Esarey, E., and Ting, A., *Phys. Rev. Lett.* **64**, 2011–2014 (1990).
8. Andreev, N.E., Gorbunov, L.M., Kirsanov, V.I., Pogosova, A.A., and Ramazashvilly, R.R., *Sov. JETP Lett.,* **55**, 571–576 (1992).
9. Antonsen, T.M. and Jr., Mora, P. *Phys. Rev. Lett.* **69**, 2204–2207 (1992).
10. Sprangle, P., Esarey, E., Krall, J., and Joyce, G., *Phys. Rev. Lett.* **69**, 2200–2203 (1992); Esarey, E., Sprangle, P., and Krall, J., *et al., Phys. Fluids.* B **5**, 2690– 2697 (1993).
11. Andreev, N.E. Gorbunov, L.M. and Kirsanov, V.I., and Pogosova, A.A., *Sov. JETP Lett.* **60**, 713–717 (1994).
12. Chuiou, T.C., Katsouleas, T., Decker, C., Mori, W.B., Wurtele, J.S., Shvets, G., and Su, J.J., *Phys. Plasmas* **2**, 310–318 (1995); Shvets, G., Wurtele, J.S., Chiou, T.C., and Katsouleas, T.C., *IEEE Trans. on Plasma Sci.* **24**, 351–362 (1996).
13. Andreev, N. E., Gorbunov, L. M., Kirsanov, V. I, Nakajima, K., and, Ogata, A., *Phys. Plasmas* **4**(3), (1997).
14. Goloviznin, V.V., van Amersfoort, P.W, Andreev, N.E., and Kirsanov, V.I., *Phys. Rev.* E **52**, 5327–5332 (1995).
15. Antonsen, T.M., Jr. and Mora, P., Phys. Fluids B **5**, 1440–1452 (1993).
16. Andreev, N.E., Gorbunov, and L.M., Kirsanov, V.I., *et al., Physica Scripta* **49** 101–109 (1994).
17. Mori, W.B., Decker, C.D., Hinkel, D.E., and Katsouleas, T., *Phys. Rev. Lett.* **72**, 1482 – 1484 (1994); Decker, C.D., Mori, W.B., Tzeng, K.-C., and Katsouleas, T., *IEEE Trans. on Plasma Sci.* **24**, 379–392 (1996).

18. Sakharov, A.S. and Kirsanov, V.I., *Phys. Rev. E* **49**, 3274–3282 (1994).
19. Esarey, E., Krall, J., and Sprangle, P., *Phys. Rev. Lett.* **72**, 2887–2890 (1994).
20. Andreev, N.E., Gorbunov, and L.M., Kirsanov, V.I. *Phys. Plasmas,* **2**, 2573–2582 (1995); Andreev, N.E., Gorbunov, L.M., and Kirsanov, V.I., *Plasma Phys. Rep.* **21**, 824–834 (1995).
21. Andreev, N.E., Gorbunov, L.M., Kirsanov, V.I.., A.A. Pogosova, and Sakharov A.A., *Plasma Phys Rep.* **22**, 379-389 (1996).
22. Andreev, N.E., Gorbunov, L.M., Kirsanov, V.I., and Sakharov, A.S., *IEEE Trans. Plasma Sci.* **24**, 363–369 (1996).
23. Andreev, N.E., Kirsanov, V.I., Sakharov, A.S., van Amersfoort, P.W., and Goloviznin, V. V., *Phys. Plasmas* **3**, 3121–3128 (1996).
24. Andreev, N.E., Kirsanov, V.I., and Sakharov A.A., *Plasma Phys Rep.* **23** (3), 277–284 (1997).

Self-Focused Particle Beam Drivers for Plasma Wakefield Accelerators

B.N. Breizman*[†], P.Z. Chebotaev,* A.M. Kudryavtsev,*
K.V. Lotov,* and A.N. Skrinsky*

*Budker Institute of Nuclear Physics 630090 Novosibirsk, Russia
[†]Institute for Fusion Studies, The Univ. of Texas at Austin, Austin, TX 78712, USA

Abstract. Strong radial forces are experienced by the particle beam that drives the wakefield in plasma-based accelerators. These forces may destroy the beam although, under proper arrangements, they can also keep it in radial equilibrium which allows the beam to maintain the wakefield over a large distance and to provide high energy gain for the accelerated particles. This paper demonstrates the existence of acceptable equilibria for the prebunched beams and addresses the issue of optimum bunch spacing, with implications for forthcoming experiments.

INTRODUCTION

The concept of plasma-based accelerators has become an area of growing interest in recent years (see, e.g., reviews[1,2] and references therein). It offers the attractive opportunity of achieving very high accelerating gradients that exceed the limits of conventional accelerators. These promising acceleration schemes involve excitation of large amplitude plasma waves by either laser pulses or relativistic charged particle beams. These waves, with phase velocity close to the speed of light, can then accelerate particles to superhigh energies in future linear colliders.

Several experiments with laser-driven plasma waves have already been initiated. Thus far, the maximum measured energy gain reported is 44 MeV.[3] The acceleration occurs in a length of 0.5 mm, which corresponds to an accelerating gradient of 90 GeV/m. More recent experiments[4] indicate even higher gradients, up to 200 GeV/m. Plasma Wakefield Acceleration (PWFA) experiments, in which particle beams are used to drive the wave, are not as numerous. Yet they are of comparable interest with the laser experiments, for both physical and engineering reasons. A successful proof-of-principal demonstration has been presented,[5-8] and now these experiments need to be developed to the stage where the accelerating gradients, which will be in the GeV/m range, are achieved over a macroscopic distance. The latter requires a better quality driving beam than those generally produced by linacs. One possible solution is to use a modulated beam from an electron-positron storage ring, as proposed in Refs. 9–11. The numerical simulations with NOVOCODE[9,12] have shown that a train of N particle bunches of moderate density $n_b \sim n_p/N$, where n_p is the plasma density, can generate the accelerating electric field E_z with amplitude up to the wave-breaking limit $E_0 = \sqrt{4\pi n_p m c^2}$.

The present paper is a part of theoretical studies related to the experiment proposed in Refs. 9–11. These studies include the analysis of plasma wave nonlinearity and address the question of whether the beam can maintain its structure in a plasma over a sufficiently long distance without being destroyed by the excited plasma waves. We drop the "rigid" driver assumption used in previous studies[9–12] and take into account the effect of plasma fields on the driver radial equilibrium. We also consider the problem of bunch sequence optimization for the case of nonlinear plasma response.

There are three different spatial scales involved in the problem. The shortest scale is the wavelength, c/ω_p, of the excited wakefield, where $\omega_p = \sqrt{4\pi n_p e^2/m}$ is the plasma frequency. Next in order is the focusing–defocusing length associated with the radial force, F_r, acting on the driving beam. This force tends to change the beam radius a over the distance $L_F \sim c\sqrt{\gamma m a/|F_r|}$, where $\gamma \gg 1$ is the relativistic factor of the driver. It is natural to choose a to be of the order of c/ω_p. A much greater value of a, at a given beam density, cannot substantially increase the accelerating gradient but it would require a beam of a much greater total current. On the other hand, having $a \ll c/\omega_p$ would make the wave excitation less efficient. With $F_r = eE_0$ and $a = c/\omega_p$, we obtain $L_F \sim \sqrt{\gamma} c/\omega_p \gg c/\omega_p$. Note that, apart from a numerical coefficient, L_F is the beta function of the driver in the plasma. The largest of the three spatial scale lengths is the deceleration length of the driving beam L_d, the distance at which beam particles lose about half of their energy. For $E_z = E_0$ we have

$$L_d \sim \frac{\gamma c}{\omega_p} \gg L_F \gg \frac{c}{\omega_p}. \tag{1}$$

The rigid beam approximation is valid only on the shortest scale length, c/ω_p. On the focusing scale length, the radial dynamics of the beam becomes important. Depending on the sign of the radial force, the particles are either confined or pushed out and lost. This loss is very fast compared to the energy loss rate. Therefore, it can be treated as instantaneous on the driver deceleration scale length. This raises the question of whether the beam can reach a radial equilibrium such that all the beam particles would experience an inward radial force while they are slowed down by the self-generated longitudinal field. This would allow the maintenance of a suitable structure of the wakefield over a distance that exceeds the focusing length. Note that such an equilibrium may in fact be reached automatically as a result of the beam self-modulation caused by the radial forces. However, it is not necessary to follow the actual radial dynamics of the beam in order to check the existence of these equilibria or to find them. What can be done instead is a procedure that modifies the beam profile by eliminating the defocused particles and calculates the fields self-consistently.

In this paper we use a modified version of NOVOCODE that includes such a procedure, and we demonstrate the existence of radial equilibria of longitudinally modulated beams. Once a suitable beam density profile is obtained from the calculations, the beam distribution function over the transverse momenta can then be found to match this profile to the shape of the radial potential well. Since there is actually no restriction on the radial profile of the beam other than being reasonably smooth and providing a potential well for the particles, one can construct solutions for which the beam distribution function is compatible with experimental constraints.

The rest of the paper is organized as follows. In Sec. II, we construct the radially self-focused density profiles for the bunched beams. Section III deals with the distribution function of beam particles. In Sec. IV, we analyze the factors that determine the optimum initial modulation of the beam. Section IV summarizes the results.

SELF-FOCUSED BUNCHED BEAM

The original version of NOVOCODE calculates the fields generated in a plasma by a rigid axisymmetric ultrarelativistic driving beam of a given current density profile $\mathbf{j}_b = q n_b c \mathbf{z}$, where q is the beam particle charge, $\mathbf{z}$ is the unit vector along the beam axis, and $n_b(r, z - ct)$ is the beam density. Unless specified otherwise, we choose the beam density to have the form

$$n_b(r, z - ct) = n_b(0, z - ct) \exp\left(-\frac{r^2}{2a^2}\right), \tag{2}$$

where $n_b(0, z - ct)$ is the beam density on the axis. The code solves the Maxwell equations

$$\operatorname{rot}\mathbf{H} = \frac{4\pi}{c}(\mathbf{j}_b - en\mathbf{v}) + \frac{1}{c}\frac{\partial \mathbf{E}}{\partial t}, \tag{3}$$

$$\operatorname{rot}\mathbf{E} = -\frac{1}{c}\frac{\partial \mathbf{H}}{\partial t}, \tag{4}$$

with the fluid equations for plasma electrons

$$\frac{\partial n}{\partial t} + \operatorname{div}(n\mathbf{v}) = 0, \tag{5}$$

$$\frac{\partial \mathbf{p}}{\partial t} + (\mathbf{v} \cdot \nabla)\mathbf{p} = -e\mathbf{E} - \frac{e}{c}[\mathbf{v} \times \mathbf{H}]. \tag{6}$$

Here $-e$ with $e > 0$ is the electron charge; n, $\mathbf{v}$, and $\mathbf{p}$ are the density, velocity, and momentum of plasma electrons, respectively; and plasma ions are treated as an immobile homogeneous background.

Equations (3)–(6) allow a traveling wave solution moving with the beam so that all quantities depend on $z - ct$ rather than on z and t separately. In this solution, the Lorentz force $q(\mathbf{E} + [\mathbf{z} \times \mathbf{H}])$ acting on a ultrarelativistic driving beam can be represented by a potential Φ as

$$q(\mathbf{E} + [\mathbf{z} \times \mathbf{H}]) = -q\nabla\Phi \qquad (7)$$

with $\Phi = 0$ at $r \to \infty$.

For an arbitrary profile of n_b, the radial dependence $\Phi(r)$ in some of the beam slices may force particles to leave the beam quickly. This typically happens on the focusing scale length, i.e., much faster than the rate at which the particles slow down.

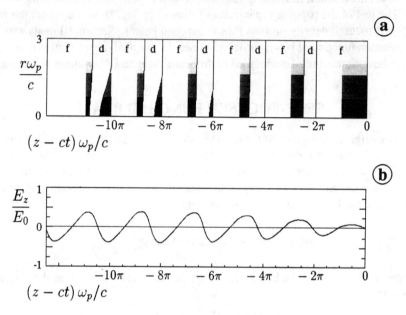

FIGURE 1. Modification of a set of equidistant rectangular bunches, due to particle defocusing by the generated plasma wave. (a) Final configuration of the beam after all defocused particles are lost. Labels f and d mark the areas of focusing and defocusing, respectively. (b) On-axis electric field generated by the modified beam.

Therefore, it is interesting to find the beam profiles such that $n_b = 0$ in the regions where the radial potential is unable to confine the particles. With this motivation, we use the following procedure to modify the beam. We first choose a set of seed bunches spaced one plasma wavelength apart so that

$$n_b(0, -x) = \begin{cases} n_0, & \text{if } 2\pi n c/\omega_p < x < (2n+1)\pi c/\omega_p, \quad n = 0, 1, \ldots; \\ 0, & \text{otherwise}. \end{cases} \qquad (8)$$

We then calculate the fields generated in the plasma by integrating Eqs. (3)–(6) in time starting from the leading edge of the beam. For every time step, we check whether the radial potential $\Phi(r)$ in the corresponding beam slice confines particles or pushes them out, and we eliminate the unconfined part of the beam slice. The resulting electric field on the beam axis and the areas occupied by the modified beam are shown in Fig. 1 (for $q < 0$).

Different shades in Fig. 1a indicate the radial fall-off of the beam density. Note that some particles of the modified beam are in the accelerating phase of the longitudinal field. These particles tend to suppress the wave rather than amplify it. In addition, the partially depleted slices make the radial profile of the beam somewhat artificial from a practical point of view. This suggests the idea of removing both the accelerated particles and the incomplete slices from the beam, which should give a stronger wakefield as well as a more natural beam shape.

In order to construct the improved beam, we start from the seed profile

$$n_b(0, -x) = \begin{cases} 0, & \text{if } x < 0, \\ n_0, & \text{if } x > 0, \end{cases} \quad (9)$$

and, while computing the fields, we totally remove the slices that contain either defocused or accelerated particles, or both.

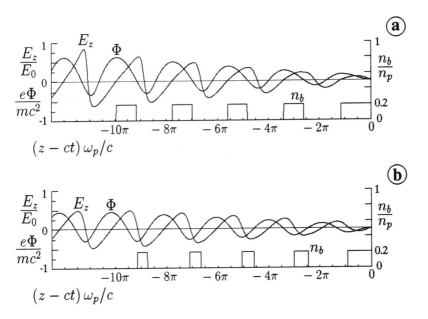

FIGURE 2. On-axis electric field E_z and potential Φ generated by chopped self-focused drivers with optimized spacing. The drivers have a Gaussian radial profile with the peak density $0.2n_p$ and the radius $a = c/\omega_p$: (a) negatively charged driver; (b) positively charged driver.

Since some of the particles removed by this procedure cannot leave the beam by themselves, the procedure implies that the beam needs to be pre-shaped by a chopping device of a kind discussed in Refs. 10 and 11. Our simulations give an example of how the beam should be chopped in order to generate a desired wakefield that keeps the beam particles focused.

The corresponding profiles for the negatively charged and the positively charged drivers are shown in Fig. 2. This figure also shows the accelerating gradient and the potential Φ on the beam axis. Note that the wakefield excited by the optimized driver (Fig. 2) is indeed stronger than that shown in Fig. 1.

The two-dimensional plot of $\Phi(r, z-ct)$ for the electron driver (Fig. 3) shows that the excited wave has a regular structure despite considerable nonlinear distortions. As the calculations continue after the last bunch of the driver, the wave eventually breaks. The larger the wave amplitude, the shorter is the interval between the last bunch and the onset of wave-breaking. For the example presented in Fig. 3, this interval is about two plasma wavelengths. Thus, the wakefield bucket next to the last driving bunch can provide acceleration and focusing for properly positioned particles.

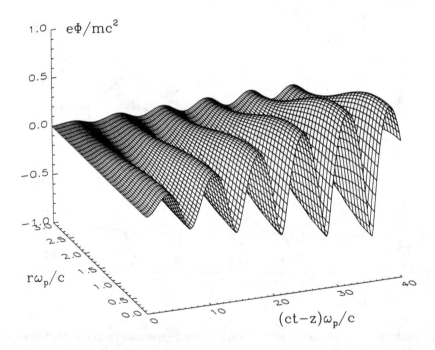

FIGURE 3. Wakefield potential $\Phi(r, z-ct)$ for the run presented in Fig. 2a.

KINETIC EQUILIBRIUM OF THE DRIVER

It has already been pointed out in Sec. I that the beam focusing/defocusing length L_F is typically much shorter than the deceleration length L_d. Therefore, the particles that are confined by the focusing field make many radial oscillations over their deceleration length. The particle orbits cross during these oscillations since different particles oscillate with different frequencies. This makes the description of the driving beam an essentially kinetic problem, whereas the fluid model still applies to the plasma. The kinetic equation for the beam distribution function f has the form

$$\frac{\partial f}{\partial t} + v_z \frac{\partial f}{\partial z} + \mathbf{v}_\perp \frac{\partial f}{\partial \mathbf{r}} + q\left(\mathbf{E} + \frac{1}{c}[\mathbf{v} \times \mathbf{H}]\right)_z \frac{\partial f}{\partial p_z} + q\left(\mathbf{E} + \frac{1}{c}[\mathbf{v} \times \mathbf{H}]\right)_\perp$$
$$\cdot \frac{\partial f}{\partial \mathbf{p}_\perp} = 0, \qquad (10)$$

where z is the coordinate along the beam axis, $\mathbf{r}$ is the two-dimensional vector perpendicular to the axis, and the subscript $\perp$ refers to the perpendicular components of the vectors.

The electric field $\mathbf{E}$ excited by an axisymmetric driver has two nonzero components, E_r and E_z; the azimuthal component of $\mathbf{E}$ vanishes. The only nonzero component of the corresponding magnetic field is H_φ. The fields E_r, E_z, and H_φ generated by a charged bunch of size c/ω_p (in both the radial and the longitudinal directions) are generally of the same order of magnitude. This means that, for a beam with a sufficiently small angular spread, one can neglect the magnetic contribution to the longitudinal force in Eq. (10).

We then introduce the function

$$F = \int f\, dp_z, \qquad (11)$$

which is the beam distribution over transverse momenta, and integrate Eq. (10) over p_z, assuming that the beam spread in p_z is sufficiently small. This integration reduces Eq. (10) to the following equation for F:

$$\frac{\partial F}{\partial t} + c\frac{\partial F}{\partial z} + \frac{c}{p_0}\mathbf{p}_\perp \cdot \frac{\partial F}{\partial \mathbf{r}} - q\nabla\Phi \cdot \frac{\partial F}{\partial \mathbf{p}_\perp} = 0, \qquad (12)$$

where p_0 is the average z-component of the particle momentum and Φ is the potential defined by Eq. (7). In this equation, we have neglected the difference between the average z-velocity of the beam and the speed of light.

We now take into account Eq. (1) and consider a z-interval that is much longer than L_F but much shorter than L_d. This allows us to construct a radial equilibrium in the cylindrical coordinate system $(r, \varphi, z - ct)$. The equilibrium distribution satisfies the equation

$$\frac{c}{p_0}\mathbf{p}_\perp \cdot \frac{\partial F}{\partial \mathbf{r}} - q\nabla\Phi \cdot \frac{\partial F}{\partial \mathbf{p}_\perp} = 0. \qquad (13)$$

It follows from Eq. (13) that the equilibrium distribution function is generally a function of two constants of motion: the perpendicular energy

$$W = \frac{cp_\perp^2}{2p_0} + q\Phi \qquad (14)$$

and the z-component of the angular momentum, $M_z = [\mathbf{r} \times \mathbf{p}]_z$. In order to simplify the problem, we assume that the distribution function has no M_z dependence and also that F vanishes for $W \geq 0$, which means that all the beam particles are trapped in the radial potential well formed by the focusing force. We now use this distribution to calculate the beam current density j_z:

$$j_z = 2\pi q p_0 \int_{q\Phi(r)}^{0} F(W)\, dW. \qquad (15)$$

In addition to Eq. (15), Φ is related to j_z by the solution of Eqs. (3)–(6). Once this additional relationship is found, subject to the constraints that all the beam particles experience radial focusing and that both j_z and Φ are monotonic functions of r, one can use Eq. (15) to obtain the equilibrium distribution function

$$F(q\Phi) = -\frac{1}{2\pi q^2 p_0} \left(\frac{\partial j_z}{\partial \Phi}\right)_{z-ct} \qquad (16)$$

This result shows the existence of many radial equilibria for longitudinally modulated beams for a broad range of beam current profiles. If applied to the optimized beam of Fig. 2 described in Sec. II, Eq. (16) indicates that the beam emittance must change along the bunch as the square root of Φ in order to keep the bunch radius constant.

Such a distribution may self-establish dynamically when the prebunched beam has a very low initial emittance. The excited wave forces the beam particles to oscillate in the radial potential well. Particle orbits in phase space tend to mix during these oscillations. This mechanism can increase the beam emittance to an equilibrium level. Depending on initial conditions, different equilibria can be reached as a result of the radial dynamics. A particle code is now under development to solve the dynamical problem, which will allow checking whether the beam can indeed relax to the described equilibrium.

Since the beam continuously loses its energy as it propagates through the plasma, its equilibrium distribution $F(W; M_z)$ has to evolve on a slow time scale. The perpendicular energy W will not be a conserved quantity in this case, but instead a radial adiabatic invariant, J, can be introduced, so that F becomes a function of J and M_z. The generalization of our results to this case is straightforward.

OPTIMUM BUNCH SPACING

The results presented in Fig. 3 show that the lengths of the oscillation cycles of the potential Φ differ from the natural period $2\pi c/\omega_p$ with some variations along the beam. We find that this deviation stems from three effects.

The first effect is independent of the driver density. It results from the system's memory about initial conditions. This can be illustrated on a simple model of a linear oscillator driven by an external force:

$$\ddot{x} + x = f(x, \dot{x}), \tag{17}$$

where

$$f(x, \dot{x}) = \begin{cases} 1, & \text{if } x > 0 \text{ and } \dot{x} > 0, \\ 0, & \text{otherwise.} \end{cases} \tag{18}$$

Equations (17) and (18) allow a straightforward analytical solution. Assuming zero initial conditions for x and $\dot{x}$, we find the following sequence of cycles for $x(t)$:

$$x_n = \begin{cases} 1 - \cos(t - t_n) + R_n \sin(t - t_n), & 0 < t - t_n < \varphi_n + \pi/2, \\ (1 + \sqrt{1 + R_n^2}) \cos(t - t_n - \varphi_n - \pi/2), & \varphi_n + \pi/2 < t - t_n < \varphi_n + 2\pi, \end{cases} \tag{19}$$

where $\varphi_n = \arctan(1/R_n)$, and t_n and R_n satisfy the following recursion equations:

$$R_0 = 0, \quad R_{n+1} = 1 + \sqrt{1 + R_n^2}, \tag{20}$$

$$t_0 = 0, \quad t_{n+1} = t_n + \arctan(1/R_n) + 2\pi. \tag{21}$$

This solution shows that every cycle is longer than the natural period 2π by the phase shift φ_n. At $n \to \infty$ the cycle length asymptotically approaches 2π. It is also interesting to note that these forced cycles accumulate a logarithmically growing total phase shift with respect to the phase of a free oscillator.

The second factor that affects the wave period is the flow of plasma electrons neutralizing the average current of the driving beam. The corresponding drift velocity can be roughly estimated as

$$|\bar{v}_z| \simeq c \frac{\bar{n}_b}{n_p}, \tag{22}$$

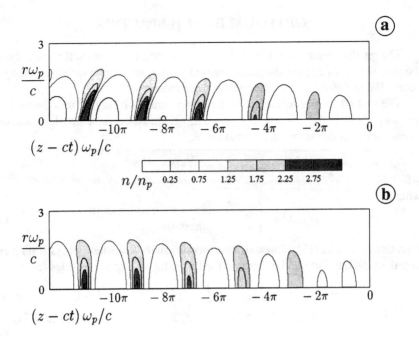

FIGURE 4. Contour plots of plasma electron density $n(r, z - ct)$ for the runs presented in Fig. 2a: (a) negatively charged driver (Fig. 2a), (b) positively charged driver (Fig. 2b).

where $\bar{n}_b$ is the average driver density and we assume $a \sim c/\omega_p$. The electron flow results in a Doppler shift of the wave frequency, which has opposite signs for positively charged and negatively charged drivers. This explains why the wave periods for the positive driver (Fig. 2b) are systematically shorter than those for the negative one (Fig. 2a). The average electron flow is also seen in the contour plots of the perturbed electron density (Fig. 4). Note that the contours have opposite tilts for positive and negative drivers. The electron drift is generally inhomogeneous, which distorts the wave profile compared to that predicted by the linear theory. A consequence of such a distortion is a non-monotonic radial profile of the potential Φ at some z-locations.

The third effect is that the wave period changes due to the relativistic reduction of the plasma frequency in the nonlinear wave.[13] This mechanism should increase the period towards the end of the driving pulse. However, in our calculations, the resulting correction to the period is roughly an order of magnitude smaller than the other two corrections.

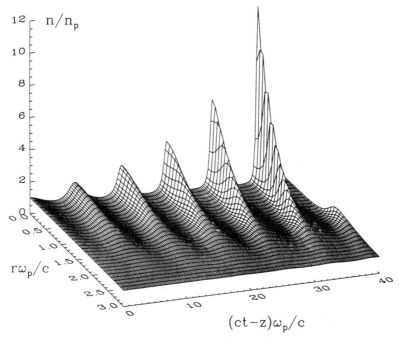

FIGURE 5. Perturbed plasma electron density $n(r, z - ct)$ for the run presented in Fig. 2a.

An interesting manifestation of wave nonlinearity is that the regions of favorable focusing for a positively charged driver are shorter than those for a negatively charged one (Fig. 2). A positive driver has to be localized within the narrow peaks of the plasma electron density in Fig. 5, while a negative driver can occupy the broad regions of density depletion. Therefore, for the same values of n_0 and a, the wake-field is somewhat stronger for a negative driver.

An important practical question from the experimental point of view is how sensitive are the amplitude and the phase of the wakefield to variations of the plasma density. In order to get an idea of this sensitivity, we take the equilibrium beam profiles presented in Fig. 2 and use them as seed profiles in the calculations with somewhat different plasma density. Similar to the run that led to Fig. 1, we modify the seed beam by eliminating the defocused particles and calculate the resulting accelerating gradient at a given distance from the leading edge of the beam. The dependence of this gradient on the plasma density is shown in Fig. 6 by solid curves. We also plot the maximum accelerating gradient after the fifth beam pulse (dotted curves). Figure 6 indicates that the phase of the excited wave is more sensitive to density variations than the wave amplitude. The asymmetry of the solid curves in

Fig. 6 suggests that it might be useful to choose the operational point for the experiment on the gentle slope of the curve rather than at the maximum. This will make the accelerating gradient less susceptible to possible errors.

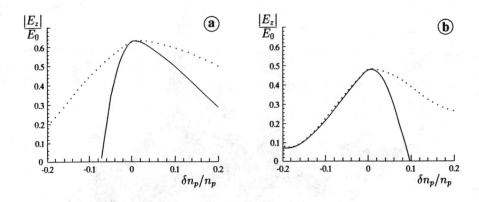

FIGURE 6. Effect of plasma density mismatch $\delta n_p/n_p$ on the wakefield generated by the optimized beams of Fig. 2. Solid curves show the accelerating gradient at a given distance after the last bunch, dotted curves the maximum gradient after the last bunch. (a) negatively charged driver (Fig. 2a), (b) positively charged driver (Fig. 2b).

SUMMARY

In this paper, we have shown that a modulated driving beam can efficiently excite nonlinear plasma waves with amplitude up to the wavebreaking limit without being destroyed by the radial defocusing forces. We have found self-consistent radial equilibria that allow the beam to propagate a distance on the order of its slowing down length. In this case, a single stage of acceleration in the beam wakefield should give an energy gain (per particle) that is roughly equal to the particle energy in the driver. For the proposed Novosibirsk plasma wakefield acceleration experiment[9-11] we expect a gain of about 0.5 GeV at the distance of 0.5 meter. Although it is conceivable that the driving beam will reach a self-modulated equilibrium state in the plasma automatically, an appropriate pre-modulation of the driver can make the outcome much more predictable, which is very important for achieving the maximum energy gain and for being able to control the fields acting on the accelerated particles. It can also be important to pre-arrange the driver emittance in accordance with the desired equilibrium profile in order to minimize uncontrollable radial oscillations. Both the modulation and the emittance adjustment can be done with the same chopper. The results presented in this paper allow optimization of the chopper design for the experiments. This optimization will take into account the beam emittance requirements and the phase shifts of the excited nonlinear wave with respect to the linear plasma wave. Our results also allow determination of a tolerable range for the plasma density variations in the experiment.

The accelerated bunch needs to be placed in the region where there is a radial potential well. Large radial forces in this region generally raise concerns about the bunch emittance. A possible way to keep the emittance low is to make the bunch radius much smaller than the wavelength (without degrading the efficiency). The length of the bunch should also be small compared to the wavelength (in order to reduce the energy spread of the accelerated particles), but it can still be much greater than the bunch radius. There can be fairly strong local perturbations of plasma electrons near this narrow bunch, which may enhance the bunch energy losses. However, a separate analysis shows that the losses cannot compete with the maximum accelerating gradient in the plasma wave as long as the total charge of the bunch is much smaller than the total charge of plasma electrons per one cubic wavelength.

Another important question is whether the equilibria found in this paper are stable with respect to nonaxisymmetric perturbations and whether similar equilibria exist for nonaxisymmetric drivers. More work is needed to address this issue quantitatively. This involves development of a hybrid code that will combine the fluid description of the background plasma with particle simulations of the beam dynamics.

ACKNOWLEDGMENTS

This work was completed during the New Ideas for Particle Accelerators program held at the Institute for Theoretical Physics, University of California at Santa Barbara. Three of us (B.N.B., K.V.L., and A.N.S.) would like to thank the program coordinator Dr. Z. Parsa and the Institute for Theoretical Physics for hosting our visit. This research was supported in part by the Russian Academy of Sciences, by the International Science Foundation grant RP 3000, by the U.S. National Science Foundation under grant number PHY94–07194, and by U.S. Department of Energy Contract No. DE-FG03-96ER-54346.

REFERENCES

1. Wurtele, J.S., *Phys. Fluids B*, 5, 7, 2363–2370 (1993).

2. Esarey, E., Sprangle, P., Krall, J. and Ting, A., *IEEE Trans. Plasma Sci.* 24, 2, 252–288 (1966).

3. A. Modena, A., Najmudin, Z., Dangor, A.E., Clayton, C.E., Marsh, K.A., Joshi, C., Malka, V., Darrow, C.B., Danson, C., Neely, D., and Walsh, F.N., *Nature* **377**, 606–608 (1995).

4. Umstadter, D., Chen, S.-Y., Maksimchuk, A., Mourou, G., and Wagner, R., *Science* **273**, 472–475 (1996).

5. Rozenzweig, J.B., Cline, D.B. *et al.*, *Phys. Rev. Lett.* **61**, 1, 98–101 (1988).

6. Nakajima, K., Enomoto, A. *et al.*, *Nucl. Instr. & Meth.* **A292**, 1, 12–20 (1990).

7. Ogata, A., "Plasma lens and wake experiments in Japan," in: *Advanced Accelerator Concepts, AIP Conference Proceedings,* ed. J.S. Wurtele, v. 279, pp. 420–449, (AIP Press, New York, 1992).

8. Berezin, A.K., Fainberg, Ya.B. *et al.*, *Plasma Phys. Reps.* **20**, 7, 596–602 (1994).

9. Breizman, B.N., Chebotaev, P.Z., *et al.*, "A Proposal for the Experimental Study of Plasma Wake-Field Acceleration at the 'BEP' Electron Storage Ring," in *Proceedings of 8th International Conference on High-Power Particle Beams,* Novosibirsk, 1990, edis. B. Breizman and B. Knyazev, vol. 1, pp. 272–279 (World Scientific, London, 1991).

10. Bechtenev, A.A., Breizman, B.N., *et al.*, "On the Possibility for Experiments on Plasma Wake-Field Acceleration in Novosibirsk," in *Advanced Accelerator Concepts, AIP Conference Proceedings,* ed. J.S. Wurtele, vol. 279, pp. 466–476, (AIP Press, New York, 1992).

11. Militsyn, B.L., Bechtenev, A.A., *et al.*, *Phys. Fluids B* **5**, 252 7, 2714–2718 (1993).

12. Breizman, B.N., Tajima, T., Fisher, D.L., and Chebotaev, P.Z., "Excitation of Nonlinear Wake Field in a Plasma for Particle Acceleration," in *Research Trends in Physics: Coherent Radiation and Particle Acceleration,* ed. A. Prokhorov, pp. 263–287 (AIP Press, New York, 1992).

13. Akhiezer A.I., and Polovin, R.V., *Sov. Phys. JETP* **3**, 696–705 (1956).

Laser Acceleration of Electrons: Zero to c in Less than Ten Microns

Donald Umstadter

Center for Ultrafast Optical Science [1]
University of Michigan,
Ann Arbor, MI 48109

Abstract. It has recently been shown that by simply focusing a high-intensity laser into a gas jet, a well-defined and low-divergence beam of relativistic electrons is produced, which is accelerated by a laser wakefield. Above a certain power threshold, the laser is observed to be relativistically self-guided, creating its own light pipe. This effectively increases the laser propagation distance (beyond the fundamental diffraction limit), decreases the electron beam divergence, and increases the electron energy.

INTRODUCTION

Due to recent advances in laser technology [1], it is now possible to generate—in the interactions of high-intensity and ultrashort-duration laser pulses with matter—the highest electromagnetic and electrostatic fields ever produced in the laboratory [2]. This has come to be known as high-field science. In terms of basic research, these interactions permit for the first time the study of optics in relativistic plasmas. Technological applications include advanced fusion energy, x-ray lasers, and table-top ultrahigh-gradient electron accelerators. We discuss several aspects that are especially pertinent to the latter.

When an intense laser enters a region of gaseous-density atoms, the atomic electrons feel the enormous laser electromagnetic field, and begin to oscillate at the laser frequency $(2\pi c/\lambda = ck)$. The oscillations can become so large that the electrons become stripped from the atoms, or ionized. At high laser intensity (I), the free electrons begin to move at close to the speed of light (c), and thus their mass m_e changes significantly compared to their rest mass.

---

[1] This work is sponsored by Department of Energy/Lawrence Livermore National Laboratory subcontract B307953 and the National Science Foundation Science and Technology Center contract PHY 8920108.

This large electron oscillation energy corresponds to gigabar laser pressure, displacing the electrons from regions of high laser intensity. Due to their much greater inertia, the ions remain stationary, providing an electrostatic restoring force. These effects cause the plasma electrons to oscillate at the plasma frequency (ω_p) after the laser pulse passes by them, creating alternating regions of net positive and negative charge, where $\omega_p = \sqrt{4\pi e^2 n_e / \gamma m_e}$, n_e is the electron density, e is the electron charge and γ is the relativistic factor associated with the electron motion transverse to the laser propagation. γ depends on the normalized vector potential, a_o, by $\gamma = \sqrt{1 + a_o^2}$, where $a_o = \gamma v_{os}/c = eE/m_o \omega c = 8.5 \times 10^{-10} \lambda[\mu m] I^{1/2}[W/cm^2]$. The resulting electrostatic wakefield plasma wave propagates at a phase velocity nearly equal to the speed of light and thus can continuously accelerate hot electrons [3]. Up to now, most experiments have been done in the self-modulated laser wakefield regime [4–6], where the laser pulse duration is much longer than the plasma period, $\tau \gg \tau_p = 2\pi/\omega_p$. In this regime, the forward Raman scattering instability can grow; where an electromagnetic wave ($\omega_o, \boldsymbol{k_o}$) decays into a plasma wave ($\omega_p, \boldsymbol{k_p}$) and electromagnetic side-bands ($\omega_o \pm \omega_p, \boldsymbol{k_o} \pm \boldsymbol{k_p}$).

RECENT RESULTS

A small number of relativistic hot electrons were observed in inertial-confinement-fusion experiments with long-pulse duration large-building size lasers and solid-density targets. However, it was shown only recently that electrons can be accelerated by a plasma wave driven by intense ultrashort-duration table-top laser pulses (I $\sim 4 \times 10^{18}$ W/cm^2, $\lambda = 1$ μm, $\tau \sim 0.5$ ps) and gaseous-density targets [7,8]. Under similar conditions, electrons were even observed to have an energies up to 44 MeV, with an energy spread of 100% [9]. It was then shown that the accelerated electron beam appeared to be naturally-collimated with a low-divergence angle (less than ten degrees), and had over 1-nC of charge per bunch [10]. Moreover, as shown in Fig. 1, acceleration occured in this experiment [10] only when the laser power exceeded a certain critical value, P_c, the threshold for relativistic self-focusing. Since then, two independent research groups have simultaneously reported direct measurements of the plasma wave amplitude with a Thomson-scattering probe pulse [11,12]. The field gradient was reported [12] to exceed that of a radio-frequency (RF) linac by four orders of magnitude ($E \geq 200$ GV/m). This acceleration gradient corresponds to an energy gain of 1 MeV in a distance of only 10 microns. The plasma wave was observed to exist for a duration of 1.5 ps or 100 plasma oscillations [12]. It was calculated that it damps only because all of the wave energy was converted to the accelerated electrons. Except for the large energy spread and low average power, these parameters compare favorably with medical linacs. In fact, the much smaller source size of a laser wakefield accelerator compared with that of a conventional linac, 10

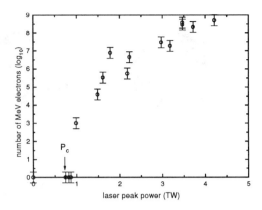

FIGURE 1. The number of relativistic electrons accelerated as a function of incident laser power focused in a gas of helium at atmospheric density.

microns compared with greater than 100 microns, may permit much greater spatial resolution for medical imaging.

This enormous field gradient would be of limited use if the length over which it could be used to accelerate electrons were just the natural diffraction length of the highly focused laser beam, which is much less than a millimeter. Fortunately, it was recently demonstrated that electrons can be accelerated beyond this distance [13]. At high laser power, the index of refraction in a plasma varies with the radius. This is both because the laser intensity varies with radius and the plasma frequency depends on the relativistic mass factor γ. Above the above-mentioned critical laser power P_c, the plasma should act like a positive lens and focus the laser beam, a process called relativistic self-focusing. This is similar to propagating a low power beam over an optical fiber optic cable, except in this case the intense laser makes its own fiber optic. The relativistically self-guided channel was found to increase the laser propagation distance by a factor of four (limited thus far only by the length of gas), decrease the electron beam divergence by a factor of two (as shown in Fig. 2), and increase the electron energy.

FUTURE DIRECTIONS

In the above experiments, the plasma itself acted as the cathode (the source of the electrons). Since the electrons were picked up by the wave and accelerated with random phases, their energy spread was large. This may be acceptable for medical radiological sources, where broadband bremstrahlung x-ray radiation is created anyway, by focusing the electron beam onto a metal target. However, in order to create monoenergetic femtosecond duration electron bunches, a new concept for laser injection of electrons was developed (as

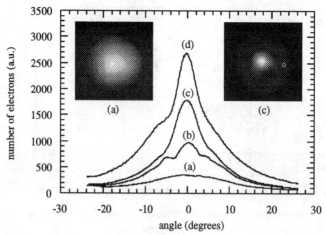

FIGURE 2. Electron beam divergence as a function of laser power. The various curves represent laser powers of $P/P_c =$ (a) 3.4, (b) 5.0, (c) 6.0, and (d) 7.5. The two insert figures show the complete beam images for curves (a) and (c).

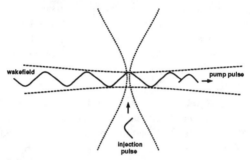

FIGURE 3. Schematic of the crossed-laser-plasma wave accelerator concept.

shown in Fig. 3), again using the plasma itself as the cathode but with laser triggering [14]. Either transverse (as shown) or collinear laser beams can be used. This is similar to giving several surfers identical pushes so that they all catch the same ocean wave in phase with one another. It was shown by use of a numerical code that with this method and by operating in the resonant wakefield regime ($\tau \sim \tau_p$), it should be possible to produce femtosecond electron bunches with energy spread at the percent level.

As an injector stage for linear electron colliders for nuclear and high-energy physics, this pulse length reduction may have several interesting consequences. In the case of an electron-electron linear collider, it would have a higher limit on attainable luminosity by permitting a shorter β (in effect, the Rayleigh parameter of the magnetic optics) at the final focus/intersection point and it would also reduce beam-beam effects by reducing the time during which the beams overlap. There would also be a reduction in beam-beam Bremsstrahlung ("beamstrahlung") due to quantum mechanical effects. Such ultrashort-duration electron bunches may also become the basis for a new generation of table-top radiation sources. They may increase both the coherence and gain of synchrotrons, free-electron lasers, or Compton-scattering sources. Additionally, the ultrashort light-pulses that they may provide might be used to as "strobes" to probe temporal dynamics on the natural timescale of important ultrafast chemical, biological and physical processes.

Another factor limiting the applicability of these laser accelerators, low average power, will be overcome as laser technology improves. With diode-pumping reaching maturity, it is expected that within the next couple of years the required laser peak intensities will be reached with 100 watt laser systems. This will be an improvement by a factor of 10^5 over the average power of the systems used in these proof-of-principle experiments, and close that of conventional RF technology.

REFERENCES

1. P. Maine, D. Strickland, P. Bado, M. Pessot and G. Mourou, *IEEE J. Quantum Electron.* **QE-24**, 398 (1988).
2. G. Mourou and D. Umstadter, *Phys. Fluids B* **4**, 2315 (1992).
3. T. Tajima and J. M. Dawson, *Phys. Rev. Lett.* **43**, 267 (1979).
4. N. E. Andreev, L. M. Gorbunov, V. I. Kirsanov, A. Pogosova and R. R. Ramazashvili, *Pis'ma Zh. Eksp. Teor. Fiz.*, **55** 551 (1992) [*JETP Lett.*, **55**, 571 (1992)].
5. T. M. Antonsen, Jr., and P. Mora, *Phys. Rev. Lett.*, **69**, 2204 (1992).
6. P. Sprangle, E. Esarey, J. Krall, and G. Joyce, *Phys. Rev. Lett.*, **69**, 2200 (1992).
7. C. A. Coverdale, C. B. Darrow, C. D. Decker, W. B. Mori, K. -C. Tzeng, K. A. Marsh, C. E. Clayton, and C. Joshi, *Phys. Rev. Lett.* **74**, 4659, (1995).
8. K. Nakajima, D. Fisher, T. Kawakubo, H. Nakanishi, A. Ogata, Y. Kato, Y. Kitagawa, R. Kodama, K. Mima, H. Shiraga, K. Suzuki, K. Yamakawa, T. Zhang, Y. Sakawa, T. Shoji, N. Yugami, M. Downer and T. Tajima, *Phys. Rev. Lett.* **74**, 4428, (1995).
9. A. Modena, Z. Najmudin, A. E. Dangor, C. E. Clayton, K. A. Marsh, C. Joshi, V. Malka, C. B. Darrow, C. Danson, D. Neely and F. N. Walsh, *Letts. Nature* **377**, 606, (1995).
10. D. Umstadter, S.-Y. Chen, A. Maksimchuk, G. Mourou, and R. Wagner, "Nonlinear Optics in Relativistic Plasmas and Laser Wakefield Acceleration of Electrons," *Science* **273**, 472 (1996).
11. A. Ting, K. Krushelnick, C. I. Moore, H. R. Burris, E. Esarey, J. Krall, and P. Sprangle, "Temporal Evolution of Self-Modulated Laser Wakefields Measured by Coherent Thomson Scattering," *Phys. Rev. Lett.* **77**, 5377 (1996).

12. S. P. Le Blanc, M. C. Downer, R. Wagner, S.-Y. Chen, A. Maksimchuk, G. Mourou and D. Umstadter, "Temporal Characterization of a Self-Modulated Laser Wakefield," *Phys. Rev. Lett.* **77**, 5381 (1996).
13. R. Wagner, S.-Y. Chen, A. Maksimchuk and D. Umstadter, "Relativistically Self-Guided Laser Wakefield Acceleration," *Phys. Rev. Lett.* (accepted for publication, 1997).
14. D. Umstadter, J. K. Kim, and E. Dodd, "Laser Injection of Ultrashort Electron Pulses into Wakefield Plasma Waves," *Phys. Rev. Lett.* **76**, 2073 (1996).

ACCELERATION AND COLLISION OF ULTRA-HIGH ENERGY PARTICLES USING CRYSTAL CHANNELS

Pisin Chen
Stanford Linear Accelerator Center *
Stanford, California 94309 USA

Robert J. Noble
Fermi National Accelerator Laboratory [†]
Batavia, Illinois 60510 USA

ABSTRACT

We assume that, independent of any near-term discoveries, the continuing goal of experimental high-energy physics (HEP) will be to achieve ultra-high center-of-mass energies early in the next century. To progress to these energies in such a brief span of time will require a radical change in accelerator and collider technology. We review some of our recent theoretical work on high-gradient acceleration of charged particles along crystal channels and the possibility of colliding them in these same strong-focusing atomic channels. An improved understanding of energy and emittance limitations in natural crystal accelerators leads to the suggestion that specially manufactured nano-accelerators may someday enable us to accelerate particles beyond 10^{18} eV with emittances limited only by the uncertainty principle of quantum mechanics.

---

*Work supported by the U.S. Department of Energy under contract No. DE-AC03-76SF00515.

[†]Work supported by the U.S. Department of Energy under contract No. DE-AC02-76CH03000.

INTRODUCTION

High-energy physics has progressed twelve orders of magnitude in energy during the last one-hundred years (1 eV to 10^{12} eV or 1 TeV). Modern high-energy colliders are both microscopes and time machines allowing us to probe fundamental physics at distances of 10^{-16} centimeters and hence understand the relevant phenomena 10^{-10} seconds after the Big Bang. It is thought by some that by advancing only one or two orders of magnitude higher in energy, experiments will place enough constraints on unified field theories to yield one consistent "Theory of Everything" including gravity. Machine builders instead assume that regardless of any intermediate discoveries, the continuing goal of experimental high-energy physics will be to achieve ultra-high center-of-mass energies in the next century. To reach these energies with their attendant high luminosity in such a brief span of time will require a radical change in accelerator and collider technology. In all likelihood more than one paradigm shift in accelerators will be needed. In this paper we review some of our recent work [1] on a concept that may enable high-energy physics to reach energies of order 10^{18} eV early in the next century.

Ten years ago the present authors made a cursory study of a concept to accelerate positively-charged particles along crystal channels by the electron plasma waves in metals [2,3]. The maximum electric field of a plasma wave is of order $\sqrt{n_o}$ V/cm, where n_o is the electron number density in units of cm^{-3}. Acceleration gradients of 100 GV/cm or more were implied based on the electron densities in solids. The strong electrostatic focusing of the atomic channels combined with the high gradients were found to maintain low beam emittance in spite of multiple scattering on channel electrons. The technological demands to excite such large amplitude plasma waves with lasers or particle beams appeared daunting then, and crystal behavior at picosecond to femtosecond time scales and high power densities was uncertain at best.

The development of ultra-short pulse-length lasers, nano-fabrication technology and a better experimental and theoretical understanding of high energy density effects in solids motivated us to return to the topic of a crystal channel accelerator. An improved picture of a crystal accelerator has emerged which allows us to further illucidate the advantages of crystals for acceleration and emittance control as well as point out the constraints imposed by the use of natural crystals as high-energy particle accelerators. Limits on high luminosity may ultimately be more difficult to overcome than achieving ultra-high energy. The quantum mechanical control of colliding particle trajectories to the level of the uncertainty principle may be required to achieve high luminosity.

CRYSTAL ACCELERATION AND EMITTANCE DAMPING

The basic concept of crystal channel acceleration combines plasma wave acceleration [4] with the well known channeling phenomenon [5] to allow positively charged particles to be accelerated over long distances without colliding with nuclei in the accelerating medium. Positively charged particles are guided by the average electric fields produced by the atomic rows or planes in a crystal. The particles make a series of glancing collisions with many atoms and execute classical oscillatory motion along the interatomic channels. In contrast negatively charged particles oscillate about the atomic nuclei and rapidly suffer large-angle Coulomb scattering.

Acceleration in the crystal is provided by an electron plasma oscillation [6] with phase velocity near the speed of light. The maximum electric field is roughly $\mathcal{E}_0 = m_e \omega_p c / e$, where m_e is the electron rest mass and $\omega_p = (4\pi n_o e^2 / m_e)^{1/2}$ is the electron plasma frequency. In convenient units, $\mathcal{E}_0(V/cm) \simeq 0.96(n_o(cm^{-3}))^{1/2}$. Doped semiconductors typically have carrier densities of 10^{14} to 10^{18} cm^{-3} corresponding to $\mathcal{E}_0 = 10$ MV/cm to 1 GV/cm, the same as typical laboratory gas plasmas. Conduction electrons in metals have densities of 10^{22} to 10^{23} cm^{-3} while the total electron density of solids is of order 10^{24} cm^{-3} implying gradients of order 1 TV/cm.

A basic obstacle to accelerating particles over long distances in crystals is beam loss from dechanneling. The transverse momentum of channeled particles increases due to collisions with electrons in the interatomic channels. Dechanneling occurs when a particle's transverse kinetic energy $E\psi^2/2$, where $E = \gamma mc^2$ is the total particle energy, allows it to overcome the channel's potential energy barrier V_c ($\sim 10 - 1000\, ze$ volts for a particle of charge ze). At this point close encounters with atomic cores quickly scatter particles out of the channel. This defines the critical channeling angle $\psi_c = (2V_c/E)^{1/2}$. In many crystals the electron density n over most of the channel is roughly constant. From Poisson's equation the channel potential energy function in either plane is simply $V = K_c x^2/2$, where $K_c = 4\pi z e^2 n$ is the focusing strength. The channel half-width a corresponds to the point where $V = V_c = K_c a^2/2$.

In the harmonic potential approximation, each crystal channel acts like a smooth focusing accelerator with betatron focusing function (wavelength/2π of transverse oscillations) $\beta_F = (E/K_c)^{1/2} = a/\psi_c$. The normalized rms channel acceptance, $A_n \equiv \gamma a^2 / 2\beta_F = \gamma a \psi_c / 2$, defines the available transverse phase space for a channeled particle. Multiple scattering in a transverse focusing system randomly excites betatron oscillations leading to growth in the normalized rms emittance $\varepsilon_n = \gamma \varepsilon = \gamma \sigma^2 / \beta_F$, where ε is the geometric emittance, and σ^2 is the rms amplitude of the particle [7]. In this terminology, dechanneling occurs when the particle emittance exceeds the channel acceptance. The emittance growth is $d\varepsilon_n/ds = (\gamma \beta_F / 2) d\langle \psi^2 \rangle_{ms}/ds$, where the increase in rms angular divergence is $d\langle \psi^2 \rangle_{ms}/ds = \psi_c^2 / 2\ell_d$. The characteristic dechanneling length is $\ell_d = \Lambda E/ze$, and Λ is the dechanneling constant ($\sim a^2$). In natural crystals where $a \simeq 1$ to 3 Å, Λ is typically of order 1 to 10 μm/MV.

Particle dechanneling in a crystal accelerator is modified by several effects. Acceleration reduces multiple scattering with increasing energy. The presence of any transverse fields in addition to the natural channel forces will change the betatron focusing function and channel acceptance by modifying the focusing strength K_c into a total strength K. Charged particles oscillating in a transverse focusing system radiate and make transitions to lower energy levels of the potential with an energy-independent decay constant $\Gamma_c = 2r_{cl}K/3mc$, where $r_{cl} = (ze)^2/mc^2$ is the classical particle radius [8,9]. These radiative transitions act to damp the particle's normalized emittance. Collisional energy loss to electrons in the channel can also damp emittance (ionization cooling) but with an energy-dependent decay parameter $E^{-1}(dE_{coll}/ds) = (E/m_ec^2)d\langle\psi^2\rangle_{ms}/ds$ for a relativistic particle.

Combining these effects, the evolution equation for the normalized emittance in a crystal channel accelerator is [1]

$$\frac{d\varepsilon_n}{ds} = -\frac{\Gamma_c}{c}(\varepsilon_n - \hbar/2mc) - \frac{zeK_ca^2}{2m_ec^2\Lambda E}\varepsilon_n + \frac{zeK_ca^2}{4mc^2\Lambda(KE)^{1/2}}, \quad (1)$$

where $E = E_i + zeGs$, E_i is the initial particle energy, and G is the (net) acceleration gradient which is assumed to be a constant. The term $\hbar/2mc$ is the minimum quantum emittance of a particle in the ground state of the transverse potential. Because of the different energy dependencies in the three terms of Eqn. (1), not all terms are equally important in an arbitrary energy regime. The effectivenes of ionization cooling clearly falls off rapidly with increasing energy. In the limit of high gradients ($G \gg K_ca^2/\Lambda m_ec^2 \sim 1$ MV/cm) and short channel lengths ($\Gamma_cs/c \ll 1$), ionization cooling and radiative damping have a negligible effect on the emittance evolution in the channel compared to acceleration, and the emittance becomes

$$\varepsilon_n = \varepsilon_{ni} + \frac{(K_c/K)A_n}{\Lambda G}(1 - (\gamma_i/\gamma)^{1/2}). \quad (2)$$

Accelerated particles remain indefinitely channeled provided $G \geq K_c/K\Lambda$. This corresponds to 1 to 10 GV/cm in natural crystal channels when $K = K_c$. Note that the equilibrium *rms* amplitude is $\sigma^2 = (K_c/K)a^2/2\Lambda G$.

For distances $s \gg c/\Gamma_c$ where radiative damping is important and $E \gg E_i$, the normalized emittance can be written approximately as

$$\varepsilon_n(s) = \frac{\hbar}{2mc} + \frac{3mc^2(K_c/K)a^2}{8ze\Lambda(zeGKs)^{1/2}}, \quad (3)$$

which damps like $\gamma^{-1/2}$ provided the net gradient G can be maintained constant with the increasing radiative energy loss. Note that the *rms* amplitude σ^2 of the channeled particle damps like γ^{-1} in this regime. The presence of K_c in Eqn. (3) reflects the deleterious effect of electron multiple scattering, and prevents one from realizing the ideal quantum emittance in such a collective accelerator.

To obtain small emittances and high luminosity in a channeling accelerator then, it is advantageous to have a high acceleration gradient and strong transverse focusing such that $K \gg K_c$. In practice the available technology will limit the plasma wave amplitude G_0 that can be generated in a crystal channel accelerator. When the magnitude of the radiative energy loss rate $(dE/ds)_{rad} = -\Gamma_c \gamma^2 K \sigma^2$ becomes comparable to zeG_0, a limiting energy is reached. In the regime $\Gamma_c s/c < 1$ where Eqn. (2) is valid, the radiation rate is proportional to γ^2, and the limit is

$$E_{max} \simeq \sqrt{\frac{3\Lambda}{zea^2 K K_c} m^2 c^4 G_0} \,. \quad (4)$$

The presence of K and K_c in Eqn. (4) reflects the competing effects of strong focusing and multiple scattering in the channel. This places a fundamental energy limit on natural crystal accelerators with $K = K_c$ because the electron density ($\sim K_c$) is fixed by the atomic structure. For example if $G_0 = 100$ GV/cm and $K = K_c = 20$ eV/Å^2, then the maximum energy is about 300 GeV for positrons, 10^4 TeV for muons and 10^6 TeV for protons. On the other hand if one can artificially arrange that $K_c < (4ze\Lambda/3a^2)K$, accelerated particles will enter the regime $s > c/\Gamma_c$ before the limit (4) is reached. Here the radiation rate is only proportional to γ, and the energy limit is $E_{max} \simeq 4m^2 c^4 \Lambda G_0/K_c a^2$.

CRYSTAL BEHAVIOR AT HIGH GRADIENTS

Only for acceleration gradients $G \geq \Lambda^{-1} \simeq 1 - 10$ GV/cm will particles remain channeled over long distances in a natural crystal accelerator. Since Λ is proportional to a^2, it may be useful to consider artificially wide channels ($a > 3$ Å) to reduce the gradient demand, at least for early experiments where gradients may be limited. Still, a large amplitude plasma wave with a field of 100 GV/cm or more is ultimately desirable to shorten the accelerator and keep the emittance as low as possible.

Two regimes of the crystal accelerator can be distinguished based on whether the plasma wave amplitude or the fields used to excite the wave are greater than or less than $I/r_a \simeq 1 - 10$ V/Å, where I is the ionization energy of electrons in an atom of size r_a. For fields greater than this, the Coulomb potential of an atom is sufficiently deformed to induce significant tunneling ionization. For an oscillating electric field $\mathcal{E}$, electrons tunnel from atoms within a time $v_e/c\epsilon\omega$, where $v_e = (2I/m_e)^{1/2}$, $\epsilon = e\mathcal{E}/m_e \omega c$ is the normalized field strength, and ω is the frequency [10]. Typically v_e/c is of order the fine-structure constant $\alpha \simeq 1/137$, so for field strengths $\epsilon > 10^{-2}$, electrons escape the atom within an oscillation period. In this high field regime, the lattice ionizes, but does not yet dissociate, on this time scale. If an intense laser ($> 10^{14} - 10^{15}$ W/cm^2) is used to build up the plasma wave, the lattice will already be in this ionized state prior to plasma-wave formation.

For laser and plasma fields below 0.1 to 1 GV/cm, reusable crystal accelerators can probably be built which might survive multiple pulses. Plasma wave decay is determined by interband transitions with a timescale of 10 to 100 ω_p^{-1} in this regime [11]. For fields above a GV/cm, only disposable accelerators, perhaps in the form of fibers or films, are possible. The lattice is highly ionized by the laser driver used to excite the plasma wave in a few optical periods, and the free electron density immediately increases to 10^{23} cm^{-3} or more for any solid. Plasma wave build-up and channeled particle acceleration must occur before the ionized lattice disrupts due to ion motion. The lattice dissociates by absorbing plasmon energy on a timescale determined by the inverse ion plasma frequency $\omega_{pi}^{-1} = (m_i/m_e)^{1/2}\omega_p^{-1} \sim 10^{-14}$ sec, where m_i is the ion rest mass. Within this time, the ions have not moved appreciably, and the lattice remains sufficiently regular to allow channeling.

The generation of large-amplitude plasma waves in a crystal requires an intense power source to supply the plasma-wave energy before the lattice dissociates. A gradient of 100 GV/cm corresponds to an energy density of 3×10^7 J/cm^3, and this must be created and used within ω_{pi}^{-1}. Conceivably side-injected laser [12], laser wakefield [13] or another mechanism could be used to excite plasma waves in a crystal channel collider. For low gradients ($<$ 1 GV/cm) reusable accelerators probably would take the form of crystal slabs on some alignable substrate. For higher gradients replaceable films or fibers are more appropriate since these are expected to be vaporized on each pulse. Alignment is certainly problematic here, and awaits the invention of fast, repeatable atomic-scale positioning. This is needed to permit staging of crystal accelerator sections with atomic precision and maintain a straight accelerator. Dislocations, unintended crystal curvature, and misalignment between sections will likely be the practical limits to long crystal accelerators.

THE CRYSTAL CHANNEL COLLIDER

The emittance solutions above suggest that small beamlets can be maintained with a high acceleration gradient and strong transverse focusing in crystal channels. As noted in Ref. 9, the small beamlets can in principle be brought into collision with a high probability if the crystals of each collider arm can be aligned channel to channel. This improves the luminosity, but limitations are still reached because the bunch population cannot be made arbitrarily high, as is true in all accelerators with small transverse dimensions and short wavelengths. The crystal lattice disrupts after about 10^{-14} sec, or a hundred plasma oscillations, so the number of accelerated bunches in each channel is limited to $n_b \simeq 100$. The number of particles in each bunch is denoted by N. The bunches pass through all bunches of the oncoming train so the luminosity is proportional to $n_b^2 N^2$. Of course the accelerating crystal contains a huge number of parallel atomic channels, n_{ch}, each accelerating its own n_b bunches. The luminosity of this parallel array of accelerators is then $L = f_{rep} n_{ch} n_b^2 N^2 \gamma / 4\pi \beta^* \varepsilon_n$. Here f_{rep} is the repetition rate of the accelerator, and β^* is the channel beta function $(E/K)^{1/2}$ since no additional focusing at the crossing is assumed.

For the sake of discussion, let us assume a natural crystal with $K = K_c$, $a \simeq 1$ Å, and that the emittance is given by Eqn. (2) with an acceleration gradient $G = 10\Lambda^{-1} \simeq 100$ GV/cm. The number of accelerated particles in each plasma oscillation bucket is limited by beam loading [14] to a value $n_{ch} N \simeq n_{ch} A_{ch} G / 8\pi e$, where $A_{ch} \simeq \pi$ Å^2 is the area of an atomic channel. This yields $N \simeq 10$, and the luminosity becomes $L(cm^{-2}sec^{-1}) \simeq 2 \times 10^{22} f_{rep} n_{ch}$. To use a proton collider for discovering new physics at a center-of-mass energy E_{cm} may require a luminosity $L(cm^{-2}sec^{-1}) \simeq 10^{29}(E_{cm}(TeV))^2$, although this may be an overestimate. This implies $f_{rep} n_{ch} \simeq 5 \times 10^{12}$ at 10^3 TeV and 5×10^{18} at 10^6 TeV. The average beam powers at these energies are 800 GW and 8×10^8 TW, respectively. These high powers result from the inherent disadvantage of having many parallel accelerators each with a small number of particles per bunch.

The situation can be improved according to Eqn. (3) by having low electron density and/or strong focusing ($K \gg K_c$) in each channel so that particles would enter the radiation damping regime where σ^2 damps like γ^{-1}, thus increasing the luminosity. The method for doing this for each channel independently is unclear, though we offer some speculation here. The desire to reach emittances limited by the uncertainty principle seems to imply the need for individually manufactured nano-accelerators bundled in a parallel array and each containing strong transverse focusing elements. The radius of each nano-accelerator tube would be much less than a plasma wavelength so a uniform electron plasma oscillation would exist in the bulk but larger than 10 Å so that the electron density near each channel center would be extremely low (to eliminate multiple scattering). Radiative damping would reduce the beam emittance in each tube to $\hbar/2mc$. This may have interesting consequences which we have only begun to explore. Such a cooled beam might exhibit a condensate behavior in which channeled particles form pairs leading to an ordered state relatively impervious to multiple scattering. In analogy to a superconducting transition this would appear to require some residual attractive force between channeled particles so that a lower energy condensed state would exist. Whether the nanotube array can exhibit a phonon spectrum suitable to produce such a residual interaction, as occurs in familiar superconductor lattices, is under study.

CONCLUSION

The concept of a crystal channel collider provides a useful arena in which to explore new ideas for particle acceleration. The chief advantages of collective acceleration in crystal channels remain the avoidance of emittance growth due to multiple scattering on atomic nuclei and the potential for very high acceleration gradients. The crystal naturally provides a confined, uniform electron plasma for acceleration and a strong focusing system to maintain a small beam size and increase luminosity. In natural crystal accelerators, multiple scattering on channel electrons competes strongly with radiative emittance damping, and keeps the transverse particle amplitudes from being reduced to the quantum mechanical limit. The resulting radiative energy loss limits the maximum attainable energy which is then proportional to the acceleration gradient that can be generated. For a gradient of 100 GV/cm, proton energies of order 10^{18} eV are possible. Channels with low electron density and/or strong additional focusing are suggested to raise the energy limit. Artificial nanotube accelerator arrays offer the possibility of cooling emittances to values determined by the uncertainty principle but still providing collective acceleration with electron plasma waves. Work in progress involves exploring whether these cooled beams might exhibit a condensate behavior which, like Cooper pairs in a superconductor, are much more impervious to scattering than single particles. This would open the way to enhancing luminosity by manipulating the wavefunctions of ultra-high energy channeled beams.

ACKNOWLEDGEMENTS

The authors wish to thank R. Carrigan, M. Downer, T. Tajima and D. Umstadter for helpful discussions.

REFERENCES

1. P. Chen and R.J. Noble, "Crystal Channel Collider: Ultra-High Energy and Luminosity in the Next Century", SLAC-PUB-7402 and Fermilab-Conf-96/441 (January 1997), in Proc. of the 7th Workshop on Advanced Accelerator Concepts, Lake Tahoe, California (October 12-18, 1996).

2. P. Chen and R. J. Noble, "A Solid State Accelerator" in *Advanced Accelerator Concepts*, AIP Conf. Proc. **156**, ed. F.E. Mills (AIP, New York, 1987), p. 222.

3. P. Chen and R. J. Noble, "Channeled Particle Acceleration by Plasma Waves in Metals" in *Relativistic Channeling*, ed. R. A. Carrigan and J. A. Ellison (Plenum, New York, 1987), p. 517.

4. For reviews, see *Advanced Accelerator Concepts*, AIP Conf. Proc. **91**, **130**, **156**, **193**, **279** and **335** (Amer. Inst. of Physics, New York).

5. For a review, see D. S. Gemmel, Rev. Mod. Phys. **46**, 129 (1974).

6. C.J. Powell and J.B. Swan, Phys. Rev. **115**, 869 (1959).

7. B.W. Montague and W. Schnell, "Multiple Scattering and Synchrotron Radiation in the Plasma Beat-Wave Accelerator", in *Laser Acceleration of Particles*, ed. C. Joshi and T. Katsouleas, AIP Conf. Proc **130** (AIP, New York, 1985),p. 146.

8. Z. Huang, P. Chen and R. Ruth, Phys. Rev. Lett. **74**, 1759 (1995).

9. P. Chen, Z. Huang and R. Ruth, "Channeling Acceleration: A Path to Ultrahigh Energy Colliders", in Proc. of the Fourth Tamura Symposium on Accelerator Physics, AIP Conf. Proc. **356**, ed. T. Tajima (AIP, New York, 1995).

10. L.V. Keldysh, Zh. Eksp. Teor. Fiz. **47**, 1945 (1964) [Soviet Physics JETP **20**, No. 5, 1307 (1965)].

11. P.C. Gibbons *et al*, Phys. Rev. **B13**, 2451 (1976). P.C. Gibbons, Phys. Rev. **B23**, 2536 (1981).

12. T. Katsouleas *et al*, IEEE Trans. Nucl. Sci. **NS-32**, 3554 (1985).

13. T. Tajima and J.M. Dawson, Phys. Rev. Lett. **43**, 267 (1979).

14. P. Chen and R. Ruth, "A Comparison of the Plasma Beat-Wave Accelerator and the Plasma Wakefield Accelerator", in *Laser Acceleration of Particles*, ed. C. Joshi and T. Katsouleas, AIP Conf. Proc. **130** (AIP, New York, 1985), p. 213.

A Survey of Microwave Inverse FEL and Inverse Cerenkov Accelerators

T.C. Marshall and T.B. Zhang*

Department of Applied Physics, Columbia University, New York NY 10025
*Omega-P, Inc., 202008 Yale Station, New Haven CT 06520

Abstract. A Microwave Inverse FEL Accelerator (MIFELA) and a Microwave Inverse Cerenkov Accelerator (MICA) are currently under construction at the Yale Beam Physics Laboratory. MIFELA and MICA will share the same injector, a thermionic cathode rf gun that should furnish 5psec, 6MeV, 0.2nC electron pulses spaced by 350psec, using microwave power of many MW provided from a 2.85GHz klystron. MIFELA is to operate with ~4MW of 11.4GHz microwave power in the TE11 mode, with beam injection into each fourth rf cycle; a variable pitch and field undulator together with a guide magnetic field are present as well. MICA will operate at 2.85GHz using an alumina-lined waveguide driven in the TM01 mode; the phase velocity is just below c, with no guide field. MIFELA produces a beam of spiralling electrons, while MICA makes an axially-directed beam. This is a survey of the operating principles of these smooth-bore "tabletop" accelerators (~ 15MeV) as they are understood prior to operation.

1. Introduction to the MICA

The stimulated Cerenkov effect is a well-understood mechanism for generating coherent radiation from an energetic electron beam [1-3]. The radiating electrons move at speed greater than that of the velocity of light in the structure (hence the name "Cerenkov"); although there are several ways to slow light waves, as a general rule the term is used when the slowing is caused by a dielectric element. When one does a linearized treatment of the fields and the self-consistent motion of the particles, a dispersion relation is obtained for growth or decay of radiation in the system. One of the three roots obtained corresponds to a damped wave; this we identify with the mechanism of stimulated absorption, whereby an electron will gain energy at the expense of the rf field. In the discussion which follows, we consider the application of stimulated absorption in the nonlinear regime of particle trapping, which applies to an electron accelerator device. This we refer to as a microwave inverse Cerenkov accelerator ("MICA").

Acceleration of the electron is done by appropriate phasing of an electron bunch which is emitted from an rf gun, so that a continuous accelerating force is applied to all the electrons, which move synchronously with the slow rf wave. Variation of the wave speed is possible by a small taper in the filling factor of the dielectric element. Thus the device resembles an rf linac, but without the periodic loading structures in the waveguide. As the MICA is smooth-bore and the motion of the particles is rather one-dimensional, we expect that the emittance of the electron beam produced will be attractive. The MICA under consideration will use a SLAC source of microwave power at 2.85GHz, and with a bunch length of only

5psec compared with the rf period of 350psec, we can expect excellent trapping and acceleration of a monoenergetic bunch of electrons. Another approach[4-7], the ICA(Inverse Cerenkov Accelerator) experiment at Brookhaven National Laboratory, uses a CO_2 laser and an axicon to accelerate an electron beam at 40MeV energy; the light wave is slowed by introducing hydrogen gas into the beamline. The gas contributes to some electron scattering, and the main disadvantage of the short laser wavelength is that electrons interact with the wave over the full range (2π) of phase; that is, the bunch length is long compared with the rf wavelength. In the MICA, the electrons move down a 1cm diameter hole in an alumina dielectric liner as a filamentary beam of under 1mm diameter. The main limitation here is that of the maximum axial field gradient (120-160kV/cm [8]) along the dielectric surface.

In this paper, we describe first the analysis of a wave inside of a dielectric annular cylinder fitted into a cylindrical waveguide. We find dispersion relations[9] for the axisymmetric TM_{01}-like mode of the system, that is, the mode that gives maximum electric field along the axis. This calculation provides both the slowing factor and the distribution of fields. We then model the acceleration and motion of electrons in the vacuum fields of this device. In this paper we shall not deal with the effects of electron beam loading (negligible for low beam current) or space charge; the reader is referred to a more comprehensive publication that includes experimental tests[10] of the breakdown fields. The objective of this effort is to determine whether a compact high quality accelerator of this type is feasible.

2. Dispersion relation and field distribution of the dielectric-loaded waveguide

We take a waveguide of radius R, lined with a dielectric sleeve, with the central vacuum hole of radius a as shown in Figure 1. Under the appropriate boundary conditions, we solve the Maxwell equations by a standard procedure[9], and arrive at a dispersion relation of the system for eigenmodes TM0n:

$$\frac{I_1(k_{1i}a)}{I_0(k_{1i}a)} = \varepsilon \frac{k_{1i}}{k_2} \frac{J_1(k_2 a) N_0(k_2 R) - J_0(k_2 R) N_1(k_2 a)}{J_0(k_2 a) N_0(k_2 R) - J_0(k_2 R) N_0(k_2 a)} \tag{1}$$

where, the functions J and N are Bessel functions and I is the modified Bessel function; $k_{1i} = -ik_1$, k_1 and k_2 are the transverse wave numbers in the vacuum hole and the dielectric element separately.

$$k_1^2 = k_0^2 - k_z^2 \tag{2}$$

$$k_2^2 = \varepsilon k_0^2 - k_z^2 \tag{3}$$

k_z is the axial wave number of the waveguide TM_{0n} mode, $k_0 = \frac{\omega}{c}$ is the wave number in free space and ε is the dielectric constant of the material. The normalized phase velocity of the mode can be obtained from the eigenvalue of Eq. (1):

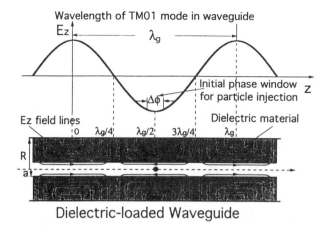

Figure 1. Longitudinal cross-section of loaded waveguide showing axial electric field distribution as well as electron phase positions along waveguide.

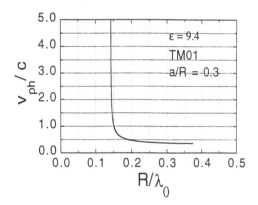

Figure 2. Normalized phase velocity v_{ph}/c vs normalized outer radius for TM_{01} mode waveguide with alumina liner. Note: $v_{ph} \leq c$ for $R \geq 0.15\, \lambda_0$.

$$\frac{v_{ph}}{c} = \frac{\omega}{k_z c} = \frac{1}{\sqrt{1 + \frac{k_{1i}^2}{k_0^2}}} \qquad (4)$$

As an example, Figure 2 shows the normalized phase velocity of TM_{01} mode as a function of the radio R/λ_0, where λ_0 is the free space wavelength of field, the dielectric constant $\varepsilon = 9.4$ for alumina, and the ratio of hole radius to outer dielectric radius is $a/R=0.30$. One sees from Figure 2 that phase velocity of c and below obtain when $R/\lambda_0 > 0.15$, which gives a relation between the waveguide radius and the rf wavelength for a slow wave structure.

With the eigenvalue of the waveguide mode solved from the dispersion relation Eq.(1), we can calculate the field distribution in this specific MICA configuration. The electromagnetic field distributions in the loaded waveguide include two separate parts: one part in the vacuum hole and the other in the dielectric, connected by the boundary conditions. For the TM_{0n} mode, only three components E_z, E_r and B_θ exist. In cylindrical coordinates, they have the following form:

when $0 \le r \le a$, in the hole:

$$E_{z1}(r,z,t) = E_0 \, I_0(k_{1i}r) \cos(\omega t - k_z z) \qquad (5.1)$$

$$E_{r1}(r,z,t) = - E_0 \frac{k_z}{k_{1i}} I_1(k_{1i}r) \sin(\omega t - k_z z) \qquad (5.2)$$

$$B_{\theta 1}(r,z,t) = - E_0 \frac{k_0}{k_{1i}} I_1(k_{1i}r) \sin(\omega t - k_z z) \qquad (5.3)$$

when $a < r \le R$, in the dielectric:

$$E_{z2}(r,z,t) = E_0 g_i \left[J_0(k_2 r) N_0(k_2 R) - J_0(k_2 R) N_0(k_2 r) \right] \cos(\omega t - k_z z) \qquad (6.1)$$

$$E_{r2}(r,z,t) = - E_0 g_i \frac{k_z}{k_2} \left[J_1(k_2 r) N_0(k_2 R) - J_0(k_2 R) N_1(k_2 r) \right] \sin(\omega t - k_z z) \qquad (6.2)$$

$$B_{\theta 2}(r,z,t) = - E_0 g_i \frac{\varepsilon k_0}{k_2} \left[J_1(k_2 r) N_0(k_2 R) - J_0(k_2 R) N_1(k_2 r) \right] \sin(\omega t - k_z z) \qquad (6.3)$$

The coefficient g_i in Eqs(6.1-6.3) is a constant relating the field amplitude in the dielectric to that in the hole:

$$g_i = \frac{I_0(k_{1i}a)}{J_0(k_2a) N_0(k_2R) - J_0(k_2R) N_0(k_2a)} \quad (7)$$

$E_0 = E_0(z)$ is axial field strength on the axis; it can be determined in terms of the total microwave power by integrating the Poyting vector in the cross-section of waveguide.

3. Single-Electron Motion and Acceleration in the Waveguide Fields

Substituting Eqs.(5.1-6.3) in the Lorentz force equation for the relativistic beam particles,

$$\frac{d\mathbf{P}}{dt} = -e\left[\mathbf{E} + \frac{1}{c}\mathbf{v} \times \mathbf{B}\right] \quad (8)$$

we are now in a position to study the motion of electrons in the MICA. In Eq.(8) the symbols have their conventional meaning; $\mathbf{P} = m\gamma\mathbf{v}$, γ is the Lorentz factor of beam particle. Since the field is not self-consistent, we account for single particle effects only; but, for a low beam current and intense driving field, this a reasonable approximation. The three components of the force equation (8) in cylindrical form are:

$$\dot{\beta}_r - \mu\beta_\theta = -\frac{e}{mc^2\gamma}\left[(1 - \beta_r^2)E_r - \beta_r\beta_z E_z - \beta_z B_\theta\right] \quad (9.1)$$

$$\dot{\beta}_\theta + \mu\beta_r = -\frac{e}{mc^2\gamma}\left[-\beta_\theta\beta_r E_r - \beta_\theta\beta_z E_z\right] \quad (9.2)$$

$$\dot{\beta}_z = -\frac{e}{mc^2\gamma}\left[-\beta_z\beta_r E_r + (1 - \beta_z^2)E_z - \beta_r B_\theta\right] \quad (9.3)$$

where, $\mu = \dot{\theta}$ is the normalized angular velocity of electron, and the overdot represents the time derivative of the quantities $d/d\tau$, $\tau = ct$.

4. Numerical results and discussion

The MICA configuration is a circular waveguide loaded by high ε dielectric material, but before selecting this configuration we also explored another dielectric structure, a rectangular waveguide loaded by a dielectric slab or slabs. No conventional TM or TE modes exist in the latter, but rather LSE(Longitudinal-Section Electric mode) or LSM(Longitudinal-Section Magnetic mode) with either zero electric or zero magnetic field normal to the dielectric surface[11]. The maximum axial field strength occurs inside the vacuum-dielectric interface and so this configuration is not appropriate for electron acceleration. However, the use of

TABLE I. Simulation parameters of MICA

Electron beam parameters
Initial electron energy $\quad\gamma_0 = 13$
Maximum initial transverse
 velocity $\quad\beta_\perp = 0.4670 \times 10^{-3}$
Initial axial velocity $\quad\beta_z = 0.9970$
Beam radius $\quad r_b = 0.05$ (cm)
$\quad r_b/R = 0.032$

Waveguide parameters
Waveguide radius $\quad R = 1.59$ (cm)
Radius of vacuum hole $\quad a = 0.48$ (cm)
$\quad a/R = 0.30$
Dielectric constant (alumina) $\quad \varepsilon = 9.4$
Waveguide length $\quad z = 150$ (cm)
Waveguide mode $\quad TM_{01}$

Radiofrequency wave
Field power $\quad P = 15$ (MW)
Maximum field strength $\quad E_{zmax} = 6.29$ (MV/m)
Frequency $\quad f_0 = 2.85$ (GHz)
Normalized phase velocity $\quad V_{ph}/c = 0.9943$
Free space wavelength $\quad \lambda_0 = 10.52$ (cm)

Waveguide wavelength $\quad \lambda_g = 10.46$ (cm)

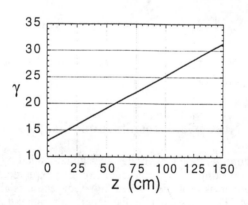

Figure 3. Electron energy as a function of the axial distance.

a high ε annulus inside circular waveguide maintains a large uniform E_z field inside the hole which is of particular interest for acceleration purposes.

The numerical simulation is based on the force Eqs.(9.1-9.3) with the field components given by Eqs.(5.1-6.3). The parameters used are listed in Table I. The electrons are taken to be monoenergetic but with small transverse velocities which vary randomly from particle to particle. At the entrance of the waveguide they are randomly distributed inside the beam cross-section as shown in Figure 4(a1). Their distribution in β_x-β_y space is shown in Figure 4(b1).

The acceleration of electrons in the MICA configuration is straightforward because of the intense axial field E_z. In Figure 3 we show the result for 15MW travelling wave power. The electron energy increases almost linearly as the particles move down the waveguide. An analytical derivation from Eqs(9.3) gives an approximate energy expression as

$$\gamma^2 = 1 + \left[\sqrt{\gamma_0^2 - 1} - \frac{eE_0\tau}{mc^2}\right]^2 \quad (10)$$

where γ_0 is the initial electron energy. When γ_0 is large(γ_0=13, for example in this case), the above equation can be further simplified as $\gamma = \gamma_0 + e|E_0|\tau/mc^2$. When $|E_0|$=210(CGS), τ=150 cm, it gives γ = 31.5, in close agreement with the simulation result. The validity of equation (10) requires that the relative phase of the electrons with respect to the rf field is π; checking the field distribution in Figure 1, one sees that this corresponds to an initial distribution of particles at the position $\lambda_g/2$. Particles in this position will experience the maximum axial field. Due to the small difference between the electron velocity and the wave phase velocity, one may expect that the electron will gradually slip from the maximum acceleration position. In the current simulation, we find a phase slippage of $\Delta\phi \approx 24°$ in one and a half meters with the electrons moving ahead of the rf field, corresponding to a slippage interval of $\Delta\tau_0 \approx 23$ps. For a rf gun with a beam bunch length of only $\Delta\tau_0 = 5$ ps, we can expect excellent trapping and acceleration of electrons during the entire propagation along the waveguide, without a taper of the dielectric element. In our calculation, the electron energy increases to about 16 MeV in 150 cm. If the dielectric surface strength is adequate, and with a higher Q structure, the electron energy can increase more.

When the electrons are located in the small "phase window" of acceleration, "momentum compaction" will retard the radial spreading out of the electrons even though the particles have an initial transverse velocity distribution. Shown in Figure 4(a2,b2) are the cross sections of the electrons in the beam at the end of the MICA in x-y and β_x-β_y space separately. Electrons remain well confined inside the hole in the dielectric and the transverse velocity spread shrinks. An algebraic analysis from Equation(9.1,9.2) gives an approximate transverse velocity as

$$\beta_\perp = \beta_{\perp 0} \exp\left[\frac{eE_0\tau}{mc^2\bar\gamma}\left(\frac{1}{2}\Delta k\tau\sin\varphi_0 + \cos\varphi_0\right)\right] \quad (11)$$

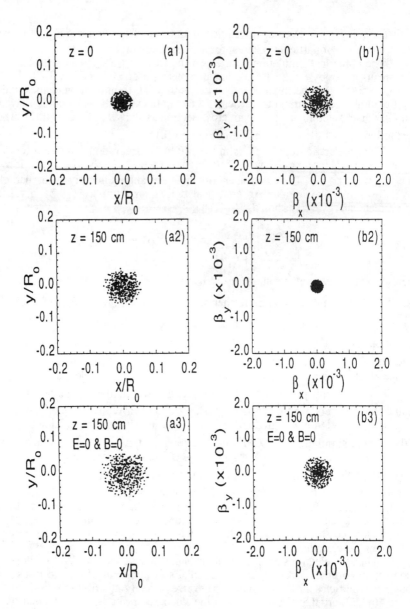

Figure 4. Beam cross-section in x-y space and β_x-β_y space: (1) at the entrance of waveguide(a1,b1); (2) at the end of waveguide with TM_{01} mode inside(a2,b2); (3) at the end of waveguide with no rf field(a3,b3). The vacuum hole radius a/R = 0.30.

where ϕ_0 is the initial phase of electrons, $\Delta k = k_z-k_0= 3.5\times10^{-3}$, and when $\phi_0 = \pi$, $\bar{\gamma}=22$, we get $\beta_\perp=0.20 \times 10^{-3}$ which is consistent with the maximum $\beta_\perp$ in Figure 4(b2). Shown in Figure 4(a3,b3) are the beam cross-section with no rf field in waveguide: in this case beam cross-section in βx-βy space does not change, whereas in x-y space it continues to spread. Beam spreading can also occur when electrons are out of acceleration phase. For instance, when particles are initially injected somewhere near $z_0 = 0$, λ_g or $\phi_0 = 0, 2\pi$, our simulation found the beam can neither be accelerated nor confined; instead the electrons will spread radially and collide with the dielectric wall.

In reference 10 is described a more detailed study of the MICA. We have found that the rf breakdown of the alumina is in excess of 8.4MV/m, and simulations with the PARMELA code find that space charge effects do not appreciably modify the conclusions we reached with regard to Figure 4.

5. Introduction to the MIFELA

The free-electron-laser has proved to be a very efficient tunable radiation source. If we regard the FEL as an electron "decelerator" where the energy of electrons is transferred in the undulator to amplify electromagnetic radiation, then it is very natural to take advantage of the analogy between FEL devices and radio frequency accelerators in which a high power electromagnetic field is used to accelerate an electron beam from low energy to high energy. The FEL operated in this process is called the inverse free-electron-laser accelerator(IFELA). The principle of the IFELA was described many years ago[12], and has been re-examined in more detail both theoretically[13,14] and experimentally[15] in the last a few years. It would generate a spiralling beam of electrons at energy ~15MeV.

The principle of acceleration is as follows. In the rest frame of the electron beam, the magnetostatic periodic field of the undulator (or wiggler) is transformed into an electromagnetic wave that beats with the microwave source; acceleration occurs by trapping a bunch of electrons into the resulting ponderomotive wave, and then increasing the velocity of this wave by tapering the undulator field and /or amplitude. We can identify some the problems facing the IFELA by considering the acceleration mechanism:

$$\frac{d\gamma}{dz} = - a_s \frac{\omega_s}{c} \frac{a_w}{\gamma} \sin \psi \qquad (12)$$

where γ is the relativistic factor of electron, $a_s=eA_s/mc^2$ is the normalized vector potential of the radiation field, ω_s is the radiation frequency, a_w/γ involves the normalized vector potential of the undulator field and amounts to the ratio of electron velocity perpendicular to the axis to the velocity parallel to the axis of the device, and is caused by the undulator interaction with the electrons. ψ is the relative phase of the electron with respect to the rf driving field.

It can be seen from (12) that the relative phase ψ plays an important part in the FEL. When $\psi > 0$, $d\gamma/dz < 0$ and this case represents stimulated emission, as in the FEL. When $\psi < 0$, then $d\gamma/dz > 0$ and this case represents stimulated absorption which is called the inverse FEL process. We can inject the electrons into a small "phase window" so that all the electrons are located in the accelerating buckets. This can be done in our example because the acceleration power is obtained from a microwave source that has a "long wavelength" (unlike some IFELA's that use a laser source of power). In our simulation the small "phase window" of phase angles is between $-\pi/8$ and $-3\pi/8$. This can be done in practice by gating the injected electron beam emerging from a rf gun and buncher cavity as it enters the IFELA; the $\pi/4$ spread corresponds to a pulse length of ~6psec. In further consideration of (12), we find that a strong rf electric field (~0.5MV/cm-- below the vacuum breakdown limit) is required, as well as a strong transverse undulator magnetic field (~0.6T), in order to achieve a sizable accelerating gradient.

The MIFELA is now under construction at Yale University. Before construction, numerical results were obtained from a nonlinear 3D FEL code "ARACHNE" [16,17]; these recent results "calibrate" reasonably well against the results of the 1D analysis presented here.

6. Theoretical Model of MIFELA

We present in this section the basic theoretical model used to describe the acceleration process in an IFELA. The system uses a helical undulator and circularly polarized driving field that propagates in a cylindrical waveguide. The fundamental equations used herein to describe the IFELA interaction are the well-known FEL equations[18,19] that define the energy and relative phase of a resonant electron in terms of the undulator magnetic field, the undulator period, and the driving field.

$$\frac{d\gamma_j}{dz} = -\frac{a_w a_s \frac{\omega_s}{c}}{\gamma_j} \sin \psi_j \left[1 - \frac{\mu^2 - 2a_w a_s \cos \psi_j}{\gamma_j^2} \right]^{-1/2} \qquad (13)$$

$$\frac{d\psi_j}{dz} = k_w + k_s - \frac{\omega_s}{c}\left[1 - \frac{\mu^2 - 2a_w a_s \cos \psi_j}{\gamma_j^2} \right]^{-1/2} + \frac{\partial \phi}{\partial z} \qquad (14)$$

Here $\mu^2 = 1 + a_w^2 + a_s^2$, z is the axial distance along the system and is the independent variable. γ_j is the relativistic factor of the jth electron, ψ_j is the phase of the jth electron with respect to the driving field, ϕ is the phase shift of the driving radiation field.

The driving field equation is a solution of the Maxwell's equations. It has the following form

$$\left[\nabla_\perp^2 + 2ik_s\frac{\partial}{\partial z} - \delta k_s^2\right] u(r, z) = -\frac{\omega_p^2}{c^2}\left\langle\frac{e^{-i(\psi - \phi)}}{\gamma}\right\rangle \quad (15)$$

$$\delta k_s^2 = k_s^2 - \frac{\omega_s^2}{c^2} \quad (16)$$

The angular brackets in the right-hand side of equation (15) indicates an ensemble average over all electrons. Equation (15) can be solved with the cylindrical waveguide boundary conditions.

Equations (13)-(15) comprise a complete set of nonlinear coupled equations. The motion equations (13) and (14) are valid for every electron, and given the initial conditions $\psi(0)$ and $\gamma(0)$ for every electron along with the parameters of the undulator, these equations can be integrated numerically to yield the value of ψ and γ as a function of the longitudinal position z. The motion of the electron in the undulator and a guiding magnetic field (used for beam transport) is represented by the term a_w/γ, and is obtained by separate solution of the orbit equation [20,21].

$$\beta_\perp = \frac{2\Omega_\perp \frac{I_1(\lambda)}{\lambda} \beta_{//}}{\Omega_0 - \gamma k_w c \beta_{//} \pm 2\Omega_\perp I_1(\lambda)} \quad (17)$$

where $\lambda = \pm \beta_\perp/\beta_{//}$, $\beta_\perp$ and $\beta_{//}$ are the transverse and axial velocities of electrons respectively; $\Omega_\perp$ and Ω_0 are the gyrofrequencies of electrons in the undulator and guiding fields; $I_1(\lambda)$ is the first-order Bessel function of imaginary argument.

The driving field, which is coupled with the equations of motion through $u(r,z) \equiv a_s e^{i\phi}$ in the right-hand of (13) and (14), is also a self-consistent solution to this set of equations. The procedure we used here is to solve the equations (13) to (15) for a large number of particles and obtain the final coordinates in phase space for every particle. The code we use is based on the equations above with a one-dimensional description for the motion of electrons, but it is two-dimensional for the driving field dynamics. This is a single-pass code corresponding to the Compton regime, for which the space charge effect is not considered: this approximation is satisfactory if the electron beam current is not higher than 100A.

7. Simulation Results and Discussion of MIFELA

Figure 5a shows the schematic of the accelerator and how the electrons are injected from an rf gun. The rf for the cavity injector is obtained from a 2.85 GHZ high power klystron; this power will be converted [22] to the fourth harmonic at 11.4 GHZ so as to drive the IFELA. We imagine that the waveguide will fill with microwaves for several nsec, interacting with a reflector in the waveguide so as to

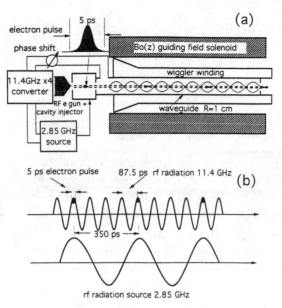

Figure 5(a): Schematic of the MIFELA;

5(b): Rf wave forms, injecting a bunch into the MIFELA.

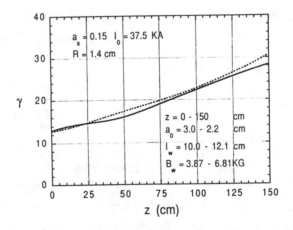

Figure 6. Electron energy as a function of the axial distance along the undulator. The dotted line is the resonant energy of trapped particles.

build up a high intensity field for electron acceleration. Then the subharmonic cavity will inject a short pulse of electrons into the IFELA, which will absorb the rf energy. The TE_{11} mode was chosen since it has a lowest cut-off frequency. In the numerical study, we increase the undulator period and the axial guiding field linearly in the accelerating section. The gradient of the axial field is determined so that the accelerator works with the stable Group I electron orbits (however, the original stability condition applies for constant energy, undulator period, and undulator field, which is not the case here).

To get the optimum accelerating gradient, we vary the undulator period $l_w(z)$ and the undulator parameter $a_w(z)$, so as to gradually increase the resonant energy of the trapped electrons and put most of the beam electrons into an accelerating bucket. Figure 6 shows the electron energy as a function of axial distance along the MIFELA. The electrons are injected monoenergetically from the gun at $\gamma = 13$ in a small initial "phase window" between $-\pi/8$ and $-3\pi/8$. The resonant energy of the design structure is shown as a dotted line. It can be seen from Fig.6 that the beam energy and the resonant energy match well along the acceleration section. The idea of small initial "phase window" is significant in this application since once injected, all the electrons are located in the accelerating buckets. Experimentally, this can be done by gating the injected electron beam emerging from a rf gun and buncher cavity as it enters the MIFELA (Figure 5b); the $\pi/4$ spread corresponds to a pulse length of ~6 psec. Figure 7 shows the phase plot and the energy spectrum of the accelerated electrons at the end of the accelerating section. Because of this small "phase window", all the electrons we simulated are trapped in the accelerating bucket. In the phase plot there is no spread of the particles, and the electron energy distribution in this case is also narrow ($\Delta\gamma/\gamma \sim 1/2 \%$).

At the end of the MIFELA, we can allow the guide and undulator fields to decrease gradually. With a sufficiently gentle gradient, one can show that the defocusing effect of the decreasing guide field can be overcome by the natural focusing of the helical undulator. If the rf field continues on in this end section of the MIFELA, we find there is still a small increase of electron energy since the electrons remain in the accelerating buckets. After that, with the decrease of guide field, a_w/γ drops, and the electrons fall out of resonance. Since the electrons have random phase with respect to the driving field, no net energy increase occurs in the end section. In this way, the output beam of the MIFELA can be extracted to zero guide field. The accelerating gradient we have obtained in this example is ~7MeV/m.

We have also designed an "entry" section of adiabatic undulator field increase for the MIFELA. A linear or sine-squared variation of the undulator field over about five periods will allow the electron beam to spiral up to a radius ~8mm, which remains approximately the same thereafter; by careful design, the electrons will enter the accelerator at the correct phase to undergo acceleration.

The device will use the power at the fourth harmonic of a 24MW, 2.856GHz XK-5 SLAC klystron as the rf pump. Calculations[22] indicate that a conversion efficiency of ~ 70% is to be expected for a cold beam, and >50% efficiency for a beam with velocity spread of 1%. We envision the use of a TE_{11m} cavity to increase the rf pump parameter in the MIFELA by approximately a factor of five compared with the value for a free travelling wave. The loaded Q would then be

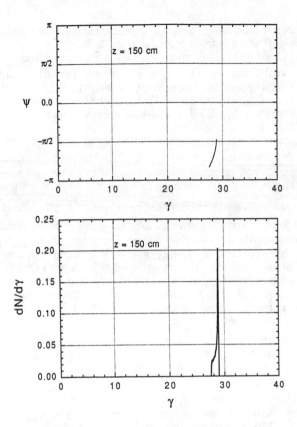

Figure 7. The electron distribution in phase space(above) and energy spectrum(below).

>5000 with an iris coupler; however the real power flow to the beam is limited to the available 12MW. This will determine the amount of charge that can be accelerated as well as the pulse repetition rate. A recent numerical study also supports a MIFELA which uses power directly obtained from the 2.85GHz klystron; this requires a larger drift tube (R = 3.1cm) and undulator period ~ 12cm together with a larger undulator winding radius, but otherwise its performance is quite similar to the higher frequency device we have just described.

Acknowledgement: This work is supported by the Department of Energy. The authors acknowledge their collaboration with J. L. Hirshfield, Rodney Yoder, and Michael LaPointe of Omega-P, Inc., and Yale University.

References

[1] J.E. Walsh, T.C. Marshall, and S.P. Schlesinger, Phys. Fluids 20, 709 (1977)
[2] J.E. Walsh and E. Fisch, Nucl. Instrum. Meth. A318 772 (1992)
[3] W. Main, R. Cherry, and E. Garate, Appl. Phys. Lett. 55, 1498 (1989)
[4] J.A. Edighoffer et al, Phys. Rev. A23 1848 (1981)
[5] J.R. Fontana and R.H. Pantell, J. Appl. Phys. 54 4285 (1983)
[6] R.D. Romea and W.D. Kimura, Phys. Rev. D42 1807 (1990)
[7] W.D. Kimura et al, Phys. Rev. Lett. 74, 546 (1995)
[8] Y. Saito, "Breakdown Phenomena in RF Windows", paper presented at Montauk, October 1994 (to be published)
[9] M. Shoucri, Phys. Fluids 26 2271 (1983)
[10] T.B. Zhang, T.C. Marshall, M.A. LaPointe, J.L. Hirshfield, and Amiram Ron, Physical Review E54, 1918 (1996)
[11] T. Moreno, "Microwave Transmission Design Data", Chapter 11, Dover, New York, [1958]
[12] R. B. Palmer, J. Appl. Phys. 43, 3014 (1972).
[13] A. C. Ting and P. Sprangle, Particle Accelerators 22, 149 (1987).
[14] S. Y. Cai and A. Bhattacharjee, Phys. Rev. A42, 4853 (1990).
[15] I. Wernick and T. C. Marshall, Phys. Rev. A46, 3566 (1992)
[16] A.K. Ganguly and H.P. Freund, Phys Rev A32, 2275 (1985)
[17] A.K. Ganguly and H.P. Freund, Phys. Flds. 31, 387 (1988)
[18] N. M. Kroll, P. L. Morton, and M. N. Rosenbluth, IEEE J. Quantum Electron. vol. QE-17, 1436 (1981).
[19] A. Bhattacharjee and S. Y. Cai, S. P. Chang, J. W. Dodd, A.Fruchtman, and T. C. Marshall, Phys. Rev. A40, 5081 (1989).
[20] L. Friedland, Phys. Fluids 23, 2376 (1980).
[21] A. K. Ganguly and H.P. Freund, Phys Rev. A32, 2275 (1985).
[22] A.K. Ganguly and J.L. Hirshfield, Phys. Rev. Lett. 70, 291 (1993)

Requirements to beam emittances at photon colliders. Laser cooling of electron beams

Valery Telnov

Institute of Nuclear Physics, 630090, Novosibirsk, Russia

Abstract. Linear colliders offer unique opportunity to study $\gamma\gamma,\gamma e$ interactions. Using the laser backscattering method one can obtain $\gamma\gamma$ and γe colliding beams with the energy and luminosity comparable to that in e^+e^- collisions or even higher (due to absence of some beam collision effects). In order to reach ultimate parameters of photon colliders the "geometric" luminosity of initial electron beams should be higher than that in e^+e^- collisions. In this report requirements to beam emittances at photon colliders are analyzed and a novel method of obtaining electron beams with small transverse emittances is considered: the electron beam is cooled during a head-on collision with the focused powerful laser pulse. With reasonable laser parameters (laser flash energy about 10 J) one can decrease the transverse normalized emittances by a factor of 10 per one stage. A limit on the final (after few stages) transverse emittance of the electron beam is much lower than that given by other known methods.

INTRODUCTION

To explore the energy region beyond LEP-II, linear colliders (LC) for the center–of–mass energy 0.5–2 TeV are developed now in the main accelerator centers [1]. In addition to e^+e^- collisions, at linear colliders one can 'convert' a electron to a high energy photon using Compton backscattering of laser light and to obtain $\gamma\gamma$ and γe collisions with energies and luminosities close to these in e^+e^- collisions [2]–[6].

To obtain high luminosity, beams in linear colliders should be very tiny. At the interaction point (IP) in the current LC designs [1], beams with transverse sizes as low as $\sigma_x/\sigma_y \sim 200/4$ nm are planned. Beams for e^+e^- collisions should be flat in order to reduce beamstrahlung energy losses. For a $\gamma\gamma$ collision beamstrahlung radiation is absent and to obtain higher luminosity electron beams with smaller σ_x can be used [4,5].

The transverse beam sizes are determined by emittances ϵ_x, and ϵ_y. The beam sizes at the interaction point (IP) are $\sigma_i = \sqrt{\epsilon_i \beta_i}$, where β_i is the beta function at the IP. With the beam energy increasing the emittance of the bunch decreases: $\epsilon_i = \epsilon_{ni}/\gamma$, where $\gamma = E/mc^2$, ϵ_{ni} is the *normalized* emittance.

The beams with a small ϵ_{ni} are usually prepared in damping rings which naturally produce bunches with $\epsilon_{ny} \ll \epsilon_{nx}$ [9]. Laser RF photoguns can also produce beams with low emittances [12]. However, for linear colliders it is desirable to have emittances even smaller than now available.

In this report a laser method of the electron beam cooling is proposed which allows further reduction of the transverse emittances after damping rings or photoguns by 1–3 orders. Using this method photon colliders can reach their maximum luminosities.

REQUIREMENTS TO BEAM EMITTANCES AT PHOTON COLLIDERS

The general scheme of a photon collider is shown in Fig. 1.

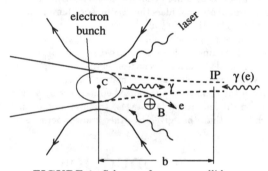

FIGURE 1. Scheme of a $\gamma\gamma$; γe collider.

Two electron beams after the final focus system are traveling toward the interaction point (IP). At a distance of about 1 cm upstream from the IP, at a conversion point (CP), the laser beam is focused and Compton backscattered by the electrons, resulting in the high energy beam of photons. With reasonable laser parameters one can "convert" most of electrons to high energy photons. The photon beam follows the original electron direction of motion with a small angular spread of order $1/\gamma$, arriving at the IP in a tight focus, where it collides with a similar opposing high energy photon beam or with an electron beam. The photon spot size at the IP may be almost equal to that of electrons at IP and therefore the luminosity of $\gamma\gamma$, γe collisions will be of the same order as the "geometric" luminosity of basic ee beams.

There are two basic collision schemes [5]:

Scheme A ("without deflection"). There is no magnetic deflection of spent electrons and all particles after the conversion region travel to the IP. The conversion point may be situated very close to the IP.

Scheme B ("with deflection"). After the conversion region particles pass through a region with a transverse magnetic field where electrons are swept aside. One thereby can achieve a more or less pure $\gamma\gamma$ or γe collisions.

In $\gamma\gamma$ collisions the minimum electron beam sizes is determined by conversion of a high energy photon into an e^+e^- pair (coherent pair creation) in the field of the opposing electron beam [7,4,5]. Coherent pair creation is exponentially suppressed when $\Upsilon = \gamma B/B_0 \leq 1$, where $B_0 = m^2c^3/e\hbar = 4.4 \cdot 10^{13}$ Gauss, $B \equiv |B|+|E|$ is the effective field of the opposing beam. But, if $\Upsilon > 1$, most of high energy photons can be converted to e^+e^- pairs during the beam collision. There are three ways to avoid this effect (i.e. to keep $\Upsilon \leq 1$):

1) to use flat beams;

2) to deflect the electron beam after conversion at a sufficiently large distance (x_0 for $E = E_0$) from the IP;

3) at certain conditions (low beam energy, long bunches) $\Upsilon < 1$ at the IP due to repulsion of electron beam [8].

Let us consider first requirements to beam sizes in the cases 1-2 (the case 3 will be considered separately). The condition $\Upsilon < 1$ corresponds approximately (at $k^2 \sim 0.4$) to the requirement [8]

$$\sigma_{x,min} \text{ or } x_0 \geq 1.5 \frac{Nr_e^2\gamma}{\alpha\sigma_z} \sim 30 \left(\frac{N}{10^{10}}\right)\frac{E_0[\text{TeV}]}{\sigma_z[mm]}, \text{ nm.} \quad (1)$$

For example, for TESLA ($N = 3.6 \cdot 10^{10}$, $\sigma_z = 0.5$ mm) and NLC ($N = 0.7 \cdot 10^{10}$, $\sigma_z = 0.1$ mm) x_0 or $\sigma_{x,min} \geq 200\ E_0[TeV]$ nm. [1]

<u>Case 1).</u> In this case the electron beam should be flat, the minimum horizontal size σ_x is given by eq. 1 and the vertical size of the electron beam should be somewhat smaller than the photon spot size arising due to the angular spread of photons in Compton scattering: $\sigma_{\gamma,y} \sim b/\gamma$, where $b \geq max(5\sigma_z, 0.1E_0[\text{TeV}]\ cm)$. The second number here is the half-length of the conversion region in the case when nonlinear QED effects are essential [5]. At $2E_0 = 500$ GeV the required beam sizes for TESLA (NLC) photon colliders are equal $\sigma_x = 50$ nm and $\sigma_y = 5$ (1) nm . For comparison, beam sizes at TESLA (NLC) planned for e^+e^- collisions are also flat and equal $\sigma_x = 800$ (320) nm and $\sigma_y = 20$ (3) nm.

<u>Case 2)</u> If the available σ_x is much smaller than $\sigma_{x,min}$ given by eq. 1, then it is reasonable to provide $\Upsilon < 1$ by deflecting a beam at the distance x_0 given by eq.(1). The minimum photon spot size in this case is

---

[1] Eq. 1 as well as eqs. 2 and 3 are approximate (due to the wide energy spectrum and repulsion during the collision) and can be used only for orientation. Final numbers should be obtained by a full simulation. These problems are considered in more details in my talk at the ITP Symposium "Future high energy colliders."

$$\sigma_{\gamma,min} \sim b/\gamma \sim \sqrt{\frac{2E_0 x_0}{eB_e}}\frac{1}{\gamma} \sim \sqrt{\frac{3Ner_e}{\alpha\sigma_z B_e}} \sim 7.5 \left(\left[\frac{N}{10^{10}}\right]\left[\frac{mm}{\sigma_z}\right]\left[\frac{T}{B_e}\right]\right)^{1/2}, \text{nm} \quad (2)$$

where B_e is the transverse magnetic field in the region between the CP and IP. For TESLA and NLC parameters $\sigma_{\gamma,min} \sim 28$ nm at $B_e = 0.5$ T. Such photon spot size can be obtained only when electron bunch sizes are smaller (by a factor 1.5–2) than these values. In this example the distance between interaction and conversion points $b = \gamma\sigma_{\gamma,min} = 1.5$ cm (due to backgrounds clear aperture (diameter) of the magnet should be larger than about 1.5–2 cm restricting a minimum length of the magnet b at the same level).

Case 3) It was noted in ref [8] that at certain conditions coherent pair creation is not essential even in a head-on collision of beams with infinitely small transverse sizes. Due to repulsion beams are separated during the collision at a rather large distance and their field at the beam axis (which influences high energy photons) is below the critical value $\Upsilon \sim 1$. This is valid for the following conditions (roughly)

$$\frac{8Nr_e^3\gamma^3}{\alpha^2\sigma_z^3} = 0.25 \left(\frac{N}{10^{10}}\right)\frac{E^3[\text{TeV}]}{\sigma_z^3[\text{mm}]} < 1 . \quad (3)$$

The maximum energy of a photon collider when beam repulsion suppresses the coherent pair creation is E = 500 GeV for TESLA ($N = 3.63 \cdot 10^{10}$, $\sigma_z = 0.5$ mm) and 200 GeV for NLC ($N = 0.7 \cdot 10^{10}$, $\sigma_z = 0.1$ mm). For NLC it is reasonable to increase a bunch length by a factor of 2 to avoid problems up to the maximum collider energy.

Below this energy the ultimate luminosity (at $z = W_{\gamma\gamma}/2E_0 > 0.65$) of these colliders can be estimated using formula

$$L_{\gamma\gamma}(z > 0.65) \sim 0.35\frac{k^2 N^2 f}{4\pi(b/\gamma)^2} \sim k^2\left(\frac{N}{10^{10}}\right)^2 \frac{E^2[\text{TeV}]f[\text{kHz}]}{b^2[\text{mm}]} \cdot 10^{36} . \quad (4)$$

For TESLA with $b = 3$ mm, $k^2 = 0.4$, f = 5.65 kHz at $E = 250$ GeV we get $L_{\gamma\gamma}(z > 0.65) \sim 2 \cdot 10^{35}$ cm^{-2}s^{-1}, providing $\sigma_x, \sigma_y \leq b/2\gamma = 3$ nm. The full simulation confirms this result. This is fantastic luminosity! May be even too large due to many background hadronic events per one bunch crossing (note that total $L_{\gamma\gamma} \sim (3-4)L_{\gamma\gamma}(z > 0.65)$). For $2E_0 = 500$ GeV the luminosity $L_{\gamma\gamma}(z > 0.65) \sim 10^{34}$ is a very good and reasonable goal. To obtain this luminosity the "geometric" ee luminosity should be larger by a factor of 10 (because $k^2 \sim 0.4$ and we consider only high energy part of the $\gamma\gamma$ luminosity spectrum.

OBTAINING OF ELECTRON BEAMS WITH LOW EMITTANCES

Present linear collider projects [1] use damping rings for obtaining beams with low emittance. Damping rings produce naturally beams with $\epsilon_{ny} \ll \epsilon_{nx}$. The attainable vertical emittances allow to produce electron beams with $\sigma_y <$ 5–10 nm at the interaction point (IP) that is sufficient for our goal. But the minimum horizontal size at the IP is too large even with the reoptimized final focus system: $\sigma_{x,min} \sim$ 300 (100) nm for TESLA and NLC respectively. As a result, these projects have $L_{\gamma\gamma}(z > 0.65) \sim (1-1.5)10^{33}$, by one order smaller than our goal. Hopefully optimization of damping rings will allow to increase the $\gamma\gamma$ luminosity by a factor of 2.

Laser RF photoguns can also produce beams with low emittance [12]. At present the normalize emittance of the best photoguns is equal $\epsilon_n = Q[nC]$ mm·mrad. With such emittances the geometric electron-electron luminosity for TESLA (NLC) is about $1(2)10^{34}$ cm^{-2}s^{-1}, as with damping rings.

As soon as $\epsilon_n \sim Q$ (in the region Q = 1–10 nC) the following further improvement is possible. Let us take many (n = 5–10) low current RF guns with low emittances. Using some energy difference one can join these bunches to one bunch (at the energy when space charge effects are small) and to get increase in the luminosity by a factor of n. (Here we assumed that the total number of accelerated particles is fixed). This scheme has not yet been considered in detail. Some visible complication in this scheme is connected with bunch compression (the joint beam has large energy spread).

Below we will consider a new option — laser cooling of electron beams, which allows to obtain normalized emittances by 3 orders of magnitude smaller than given by photoguns. This is much better than necessary for solving problems with the luminosity at photon colliders.

LASER COOLING OF ELECTRON BEAMS FOR LINEAR COLLIDERS

Principle

The idea of laser cooling of electron beams is very simple. During a collision with laser photons (in the case of the strong field it is more correct to consider the interaction of an electron with an electromagnetic wave) the transverse distribution of the electrons (σ_i) remains almost unchanged. The angular spread (σ_i') is also almost constant because the scattered photons follow along initial electron trajectories with a small additional spread. So, the emittance $\epsilon_i = \sigma_i \sigma_i'$ remains almost unchanged. At the same time the energy of electrons decreases from E_0 to E. This means that the transverse normalized emittances have decreased: $\epsilon_n = \gamma\epsilon = \epsilon_{n0}(E/E_0)$.

One can reaccelerate the electron beam up to the initial energy and repeat the procedure. Then after n stages of cooling $\epsilon_n/\epsilon_{n0} = (E/E_0)^n$ (if ϵ_n is far from its limit).

In this method we have to consider first the following problems: 1) requirements on laser parameters (these parameters should be attainable); 2) energy spread of the beam after cooling (at the final energy of a linear collider it is necessary to have $\sigma_E/E \sim 0.1\%$; also with a large energy spread it is difficult to repeat cooling many times due to the problem of beam focusing); 3) limit on the final normalized emittances (it is desirable to have this limit lower than that obtained with storage rings and photoguns); 4) depolarization of electron beams (polarization is very important for linear colliders).

Laser parameters

In the cooling region a laser photon with the energy ω_0 (wave length λ) collides almost head–on with an electron with the energy E. The kinematics is determined by two parameters x and ξ [3–5]. The first one

$$x = \frac{4E\omega_0}{m^2c^4} = 0.019 \left[\frac{E}{\text{GeV}}\right]\left[\frac{\mu\text{m}}{\lambda}\right] \tag{5}$$

determines the maximum energy of the scattered photons: $\omega_m = Ex/(x+1) \sim 4\gamma^2\omega_0$ ($x \ll 1$). If the electron beam is cooled at the initial energy $E_0 = 5$ GeV (after damping ring and bunch compression) and $\lambda = 0.5$ μm (Nd:glass laser) then $x_0 \simeq 0.2$. The second parameter is

$$\xi = \frac{eF_0\hbar}{m\omega_0 c}, \tag{6}$$

where F_0 is the field strength (E_0, B_0). At $\xi^2 \ll 1$ an electron interacts with one photon from the field (Compton scattering, undulator radiation), while at $\xi^2 \gg 1$ an electron scatters on many laser photons simultaneously (synchrotron radiation (SR), wiggler). We will see that in the considered method ξ^2 may be "small" and "large".

In the cooling region near the laser focus the r.m.s radius of the laser beam depends on the distance z to the focus (along the beam) in the following way [3]: $r_\gamma = a_\gamma\sqrt{1 + z^2/\beta_\gamma^2}$, where $\beta_\gamma = 2\pi a_\gamma^2/\lambda$, a_γ is the r.m.s. focal spot radius. The density of laser photons $n_\gamma = (A/\pi r_\gamma^2\omega_0)\exp(-r^2/r_\gamma^2)F_\gamma(z+ct)$, where A is the laser flash energy and $\int F_\gamma(z)dz = 1$.

In the case of strong field ($\xi^2 \gg 1$) it is more appropriate to speak in terms of strength of the electromagnetic field which is $\bar{B}^2/4\pi = n_\gamma\omega_0$, $B = B_0\cos(\omega_0 t/\hbar - kz)$. Assuming $F_\gamma = 1/l_\gamma$ and $\beta_\gamma \ll l_\gamma \simeq l_e$ we obtain the ratio of emittances before and after the laser target

$$\frac{\epsilon_{n0}}{\epsilon_n} \simeq \frac{E_0}{E} = 1 + \frac{r_e^2}{3m^2c^4}\int B_0^2 dz = 1 + \frac{64\pi^2 r_e^2 \gamma_0}{3mc^2 \lambda l_e}A \qquad (7)$$

$$A[J] = \frac{25\lambda[\mu m\,]l_e[\text{mm}\,]}{E_0[\text{GeV}\,]}\left(\frac{E_0}{E} - 1\right). \qquad (8)$$

These equations are correct at $x \ll 1$ for any value of ξ^2. For example: at $\lambda = 0.5\ \mu m$, $l_e = 0.2$ mm, $E_0 = 5$ GeV, $E_0/E = 10$ the required laser flash energy $A = 4.5$ J. To reduce the laser flash energy in the case of long electron bunch one can compress the bunch (length) before cooling as much as possible and after cooling to stretch it up to the required value.

Undulator and wiggler regimes

The eqs (7,8) were obtained for $\beta_\gamma \ll l_\gamma \sim l_e$ and give the minimum flash energy for the certain E_0/E ratio. For the further estimation of the photon density at the laser focus we will assume $\beta_\gamma \sim 0.25 l_e$. In this case the required flash energy is still close to the minimum one, but the field strength is not so high as for very small β_γ. From previous equation for $\beta_\gamma = 0.25 l_e$ it follows $B_0^2/(8\pi) = \omega_0 n_\gamma = A/(\pi a_\gamma^2 l_e) = 8A/(\lambda l_e^2)$. Substituting B_0 to (6) we get

$$\xi^2 = \frac{16 r_e \lambda A}{\pi l_e^2 mc^2} = \frac{3\lambda^2}{4\pi^3 r_e l_e \gamma_0}\left(\frac{E_0}{E} - 1\right) =$$

$$= 4.3 \frac{\lambda^2[\mu m\,]}{l_e[\text{mm}\,]E_0[\text{GeV}\,]}\left(\frac{E_0}{E} - 1\right). \qquad (9)$$

Example: for $\lambda = 0.5\ \mu m$, $E_0 = 5$ GeV, $E_0/E = 10$, $l_e = 0.2$ mm (the NLC project) $\Rightarrow \xi^2 = 9.7$. For larger bunch lengths and shorter wave lengths, ξ^2 may be smaller. So, both "undulator" and "wiggler" cases are possible.

Later we will see that in order to have lower limit on emittance and smaller depolarization it is necessary to have a low ξ^2. With a usual optics one can reduce ξ^2 only by increasing l_γ (and β_γ) with simultaneous increasing the laser flash energy. From (8) and (9) we get

$$A \propto \frac{\lambda^3}{\gamma_0^2 \xi^2}\left(\frac{E_0}{E} - 1\right)^2. \qquad (10)$$

Stretching of focus depth

Is it possible to reduce ξ^2 keeping all other parameters (including flash energy) constant? Yes, providing a way to stretch the focus depth without changing the radius of this area is found. In this case the collision probability (or $\int B^2 dz$) remains the same but maximum value of ξ^2 will be smaller. With a usual lens (focusing mirror) this is impossible, but it seems that this problem can be solved using the nonmonochromaticity of laser light together with a chirped pulse technique (explanation is below).

Let us put somewhere on the way of the laser beam a lens with dispersion, then the rays with the different wave lengths will be focused at different distances from the final focusing mirror. However this is not sufficient for our task. We want to have the scheme where the short electron bunch collides on its way sequentially with n light pulses of approximately the same length $l_\gamma \sim l_e$ and focused with $2\beta_\gamma \sim l_e$. Furthermore, each laser subbunch should come to its focal point exactly in the moment when the electron bunch cross this area. However, if the short laser bunch is focused by a dispersive focusing system, then the rays which are focused closer to the focusing mirror will come to their focus earlier, while the electron bunch moving towards the mirror should first collide with the rays which are focused further from the mirror (just opposite to our desire). This problem is solved in a natural way using a *chirped pulse technique* ("chirped pulse" means a pulse with time-frequency correlation) [10,11]. Namely this technique is used now for obtaining very powerful short laser pulses. Chirped pulse are obtained from a short (with large bandwidth) laser pulses using a grating pair. After passing the grating pair the short "white" pulse becomes long and chirped. The long pulse is amplified without problem of nonlinear effects in some media and then is compressed by the similar method to the short bunch. Using this wonderful technique one can prepare the chirped laser bunch of the necessary length which (after dispersive element) can be focused onto the electron beam in many focal points (stretched focal area) with necessary time delays.

In ref. [11] the authors have considered a laser scheme where one short laser bunch is splited by the (dispersive) grating into twenty separate lines and after amplification all twenty bunches are joined together with the help of the other grating. It is obvious that one can prefocus separate beams by the mirrors with the somewhat different focal lengths, so that the joined beam will be focused by the final (nondispersive) lens to many focal points.

The number n (the length of focal region/$2\beta_\gamma$, where $\beta_\gamma \sim 0.5 l_e$) depends on the stretching factor we want to get. There is one principal restriction on n: along cooling length $L \approx n \cdot l_e$ the transverse size of the electron beam should be smaller than the laser spot size $a_\gamma \simeq \sqrt{\lambda \beta_\gamma / 2\pi} \sim \sqrt{\lambda l_e / 4\pi}$. In further examples we will use $n \sim 10$ for stretching the cooling region from 100 μm to 1 mm. Using larger n will only improve the quality of the cooled electron beam (especially polarization).

Energy spread of cooled electrons

The energy spread of electrons after cooling which arises due to a quantum-statistical nature of the radiation. After losses of energy ΔE, the increase of the energy spread $\Delta(\sigma_E^2) = \int_0^\infty \varepsilon^2 \dot{n}(\varepsilon) d\varepsilon = -aE^2\Delta E$, where $a = 8\omega_0/3m^2c^4 = 2x_0/3E_0$ for the Compton case and $a = 55\hbar e B_0/(8\pi\sqrt{3}m^3c^5)$ for the "wiggler" case [9].

There is the second effect which leads to decreasing of the energy spread. It is due to the fact that $dE/dx \propto E^2$ and electron with higher (lower) energy than average one loses more (less) than on average. This results in the damping: $d(\sigma_E^2)/\sigma_E^2 = 4dE/E$ (here dE has the negative sign). The full equation for the energy spread: $d\sigma_E^2 = -aE^2 dE + 4(dE/E)\sigma_E^2$, with a solution

$$\frac{\sigma_E^2}{E^2} = \frac{\sigma_{E_0}^2 E^2}{E_0^4} + aE_0 \frac{E}{E_0}\left(1 - \frac{E}{E_0}\right) \sim$$

$$\sim \frac{\sigma_{E_0}^2 E^2}{E_0^4} + \frac{2}{3}x_0(1 + \frac{55\sqrt{3}}{64\pi}\xi)\frac{E}{E_0}\left(1 - \frac{E}{E_0}\right). \qquad (11)$$

Here the result for the Compton scattering and SR are joined together. Example: at $\lambda = 0.5$ μm, $E_0 = 5$ GeV ($x_0 = 0.19$) and $E_0/E = 10$ only the first Compton term gives $\sigma_E/E \sim 0.11$ and with the second term ($\xi^2 = 9.7$, see the example above) $\sigma_E/E \sim 0.17$.

What σ_E/E is acceptable? In the last example $\sigma_E/E \sim 0.17$ at E = 0.5 GeV. This means that at the collider energy E = 250 GeV we will have $\sigma_E/E \sim 0.034\%$, that is better than necessary (about 0.1 %).

In a two stage cooling system, after the reacceleration to the initial energy $E_0 = 5$ GeV the energy spread $\sigma_E/E_0 \sim 1.7\%$. For this value there may be a problem with focusing of electrons which can be solved using the focusing scheme with correction of the chromatic abberations. What are the resources if a smaller energy spread is necessary? After the reacceleration to the initial energy $\sigma_E/E_0 = (\sigma_E/E)(E/E_0)$. One can find that the first (Compton) term $\sigma_E/E_0 \propto (E_0/\lambda)^{1/2}(E/E_0)^{3/2}$; the second (SR) $\propto (E_0/l_e)^{1/4}(E/E_0)^{5/4}$ for $l_\gamma \sim l_e$ and $\propto \lambda^{1/4}(E/E_0)/A^{1/4}$ for free A (and $l_\gamma \sim \beta_\gamma > l_e$). Stretching of the cooling region (as it was discussed above) also helps: $\sigma_E/E_0 \propto 1/n^{1/4}$ (only the second term).

Resume: the energy spread in the one stage cooling scheme is not a problem; for the multistage cooling system one has to use the special focusing system with chromatic corrections in front of each next stage.

Minimum emittance

Case of Compton scattering

Minimum normalized emittance is determined by the quantum nature of the radiation. Let us start with the case of pure Compton scattering at $\xi^2 \ll 1$ and $x_0 \ll 1$. In this case the scattered photons have the uniform energy distribution: $dp = d\omega/\omega_m$, where $\omega_m = 4\omega_0\gamma^2$. The angle of the electron after scattering [3] $\theta_1^2 = (\omega_m\omega - \omega^2)/(\gamma^2 E^2)$. After averaging over the energy spectrum we get the average θ_1^2 in one collision: $\langle\theta_1^2\rangle = 8\omega_0^2/(3m^2c^4)$. After many collisions the r.m.s. angular spread in i=x,y projection $\Delta\langle\theta_i^2\rangle = 0.5\Delta\langle\theta^2\rangle = 0.5n\langle\theta_1^2\rangle = -0.5(\Delta E/\bar{\omega})\langle\theta_1^2\rangle = -\omega_0\Delta E/3E^2$.

The normalized emittance $\epsilon_{ni}{}^2 = (E^2/m^2c^4)\langle r_i^2\rangle\langle\theta_i^2\rangle$ does not change when $\Delta\langle\theta_i^2\rangle/\langle\theta_i^2\rangle = -2\Delta E/E$. Substituting $\langle\theta_i^2\rangle$ and taking into account that $\langle\theta_i^2\rangle \equiv \epsilon_{ni}/\gamma\beta_i$ we get the equilibrium emittance due to the Compton scattering

$$\epsilon_{ni,min} \approx 0.5\gamma E\beta\Delta\langle\theta_i^2\rangle/\Delta E = \frac{\omega_0}{6mc^2}\beta_i = \frac{\pi}{3}\frac{\lambda_C}{\lambda}\beta_i =$$

$$= \frac{4 \cdot 10^{-8}\beta_i[mm]}{\lambda[\mu m]} \; cm \cdot rad, \qquad (12)$$

where $\lambda_C = \hbar/mc$. For example: $\lambda = 0.5\ \mu m, \beta = l_e/2 = 0.1$ mm $(NLC) \Rightarrow \epsilon_{n,min} = 0.8 \cdot 10^{-8}$ cm·rad. For comparison in the NLC project the damping rings have $\epsilon_{nx} = 3 \cdot 10^{-4}$ cm·rad, $\epsilon_{ny} = 3 \cdot 10^{-6}$ cm·rad.

Minimum normalized emittance in the wiggler regime

Let us consider now the case $\xi^2 \gg 1$ when the electron moves as in the wiggler. Assume that the wiggler is planar and deflects the electron in the horizontal plane. If the electron with the energy E emits the photon with the energy ω along its trajectory the emittance changes as follows [9]: $\delta\epsilon_x = (\omega^2/2E^2)H(s)$; $H(s) = \beta_x\eta_x'^2 + 2\alpha_x\eta_x\eta_x' + \gamma_x\eta_x^2$; where $\alpha_x = -\beta_x'/2$, $\gamma_x = (1 + \alpha_x^2)/\beta_x$, β_x is the horizontal beta-function, η_x is the dispersion function, s is the coordinate along the trajectory. For $\beta_x = const$ the second term H is equal to zero, the second term in the wiggler with $\lambda_w \ll \beta$ is small, so with a good approximation $H(s)=\beta\eta'^2$. In the sinusoidal wiggler field $B(z) = B_w\cos k_w z$, $k_w = 2\pi/\lambda_w$, $\eta'' = 1/\rho$ one can find that $\eta' = (eB_w/k_w E)\sin k_w z$. An increase of ϵ_x on the distance dz

$$\Delta\epsilon_x = \int \frac{H}{2}\left(\frac{\omega}{E}\right)^2 \dot{n}(\omega)d\omega dt = \frac{55}{48\sqrt{3}}\frac{r_e\hbar c}{(mc^2)^6}E^5\langle\frac{H}{\rho^3}\rangle dz,$$

where $\langle H/\rho^3\rangle_w = 8\beta_x\lambda_w^2(eB_w)^5/(140E^5\pi^3)$ for the wiggler. Energy losses averaged over the wiggler period $\Delta E = r_e^2 B_w^2 E^2 dz/(3m^2c^4)$. The normalized emittance $\epsilon_n = \gamma\epsilon$ is not changed when $E d\epsilon + \epsilon\, dE = 0$. Using the obtained equations and replacing B_w by $2B_0$, λ_w by $\lambda/2$ we obtain the equilibrium normalized emittance in the linear polarized electromagnetic wave for $\xi^2 \gg 1$

$$\epsilon_{nx} = \frac{11 e^3 \hbar c \lambda^2 B_0^3 \beta_x}{24\sqrt{3}\pi^3 (mc^2)^4} = \frac{11}{3\sqrt{3}} \frac{\lambda_C}{\lambda} \beta_x \xi^3 \approx$$

$$\approx \frac{8 \cdot 10^{-8} \beta_x [\text{mm}] \xi^3}{\lambda[\mu\text{m}]} \quad \text{cm·rad.} \tag{13}$$

Using eq.(9) we can get a scaling of minimum ϵ_{nx} for a multistage cooling system with a cooling factor E_0/E in one stage: $\epsilon_{nx} \propto \beta_x \lambda^2 (E_0/E)^{3/2}/(l_e\gamma_0)^{3/2}$ when $l_\gamma \sim l_e$(minimum A) and $\epsilon_{nx} \propto \beta_x \lambda^{7/2}(E_0/E)^3/(\gamma_0^3 A^{3/2})$ for free A and $l_\gamma > l_e$ (for $\beta_x = const$). Stretching the laser focus depth by a factor n one can further reduce the horizontal normalized emittance: $\epsilon_{nx} \propto 1/n^{1/2}$(if $\beta_x \propto n$). For our previous example we have $\xi^2 = 9.7$ and $\epsilon_{nx} = 5 \cdot 10^{-7}$ cm·rad (in the NLC $\epsilon_{nx} = 3 \cdot 10^{-4}$ cm·rad). Stretching cooling region with $n = 10$ further decreases the horizontal emittance by a factor 3.2.

Comparing with the Compton case (12) we see that in the strong electromagnetic field the horizontal emittance is larger by a factor ξ^3. The origin of this factor is clear: $\epsilon_{nx} \propto \eta_x'^2 \omega_{crit.}$, where $\eta_x' \sim \xi\theta_{compt.}$ and $\omega_{crit.} \sim \xi\omega_{compt.}$.

Let us estimate roughly the minimum vertical normalized emittance at $\xi \gg 1$. Assuming that all photons are emitted at the angle $\theta_y = 1/(\sqrt{2}\gamma)$ with the $\omega = \omega_c$ similarly to the Compton case one get $\Delta\langle\theta_y^2\rangle = (\omega_c \Delta E)/(2\gamma^2 E^2) = -(3e\hbar \bar{B}_w \Delta E)/(4E^2 mc)$. Using the first part of eq.(12) we get minimum vertical normalized emittance for $\xi \gg 1$

$$\epsilon_{ny\,min} \sim \frac{3}{8} \frac{\hbar e \bar{B}_w}{m^2 c^3} \beta_y = \frac{3}{2\pi} \frac{\hbar e \bar{B}_0}{m^2 c^3} \beta_y = 3 \left(\frac{\lambda_C}{\lambda}\right) \beta_y \xi \approx$$

$$\approx \frac{1.2 \cdot 10^{-7} \beta_y [\text{mm}] \xi}{\lambda[\mu\text{m}]} \quad \text{cm·rad.} \tag{14}$$

For the previous example (NLC beams) eq.(14) gives $\epsilon_{ny\,min} \sim 7.5 \cdot 10^{-8}$ cm·rad (for comparison in the NLC project $\epsilon_{ny} = 3 \cdot 10^{-6}$ cm·rad). The scaling: $\epsilon_{ny} \propto \beta_y(E_0/E)^{1/2}/(l_e\gamma_0)^{1/2}$ when $l_\gamma \sim l_e$(minimum A) and $\epsilon_{ny} \propto \beta_y \lambda^{1/2}(E_0/E)/(\gamma_0 A^{1/2})$ for free A and $l_\gamma > l_e$.

For arbitrary ξ the minimum emittances can be estimated as the sum of (12) and (13) for ϵ_{nx} and sum of (12) and (14) for ϵ_{ny}

$$\epsilon_{nx} \approx \frac{\pi}{3} \frac{\lambda_C}{\lambda} \beta_x (1 + 2\xi^3); \quad \epsilon_{ny} \sim \frac{\pi}{3} \frac{\lambda_C}{\lambda} \beta_y (1 + 3\xi). \tag{15}$$

Depolarization

Finally, let us consider the problem of the depolarization. For the Compton scattering the probability of spin flip in one collision is $w = (3/40)x^2$ for $x \ll 1$ (it follows from formulae of ref. [14]). The average energy losses in one collision are $\bar{\omega} = 0.5xE$. The decrease of polarization degree after many collisions $dp = 2wdE/\bar{\omega} = (3/10)x(dE/E) = (3/10)x_0(dE/E_0)$. After integration we obtain the relative decrease of longitudinal polarization ζ during one stage of the cooling (at $E_0/E \gg 1$)

$$\Delta\zeta/\zeta = 0.3x_0 \propto E_0/\lambda, \qquad (16)$$

For $\lambda = 0.5$ μm and $E_0 = 5$ GeV we have $x_0 = 0.19$ and $\Delta\zeta/\zeta = 5.7\%$. This is valid only for $\xi^2 \ll 1$. In the case of strong field ($\xi^2 \gg 1$) the spin flip probability per unit time is the same as in the uniform magnetic field [13] $w = (35\sqrt{3}r_e^3\gamma^2 ce\bar{B}^3)/(144\alpha(mc^2)^2)$, where for the wiggler $\bar{B}^3 = (4/3\pi)B_w^3$. Using the relation between dE and dt in the wiggler we get

$$\frac{\Delta\zeta}{\zeta} = \int \frac{35\sqrt{3}er_e B_0}{9\pi\alpha(mc^2)^2} dE \sim \frac{35\sqrt{3}}{36\pi} x_0 \xi. \qquad (17)$$

For the general case the depolarization can be estimated as the sum of equations (16) and (17)

$$\Delta\zeta/\zeta = 0.3x_0(1 + 1.8\xi). \qquad (18)$$

For the previous example with $\xi^2 = 9.7$ and $x_0 = 0.19$ we get $\Delta\zeta/\zeta = 0.057 + 0.32 = 0.38$, that is not acceptable. This example shows that the depolarization effect imposes the most demanding requirements on parameters of the cooling system. The main contribution to depolarization gives the second term. One can decrease ξ by increasing l_γ and β_γ. In this method the required flash energy increases and attainable ξ depends on available laser flash energy. From (18) and (10) we can get scaling for the second term $\Delta\zeta/\zeta \propto \lambda^{1/2}(E_0/E-1)/A^{1/2}$. Another method is stretching of the focus depth, which does not require increasing laser flash energy. Stretching by a factor n reduces the second term as $1/\sqrt{n}$. After stretching the cooling region by a factor n=10 we get $\Delta\zeta/\zeta = 0.057 + 0.1 \sim 15\%$.

CONCLUDING REMARKS

Possible sets of parameters for the laser cooling: $E_0 = 4.5$ GeV, $l_e = 0.2$ mm, $\lambda = 0.5$ μm, flash energy $A \sim 5 - 10$ J, focusing system with stretching factor n=10. Final electron bunch will have the energy 0.45 GeV with the energy spread $\sigma_E/E \sim 13\%$, the normalized emittances $\epsilon_{nx}, \epsilon_{ny}$ are reduced by a factor 10, limit on the final emittance $\epsilon_{nx} \sim \epsilon_{ny} \sim 2 \cdot 10^{-7}$ cm·rad at $\beta_i = 1$ mm, depolarization $\Delta\zeta/\zeta \sim 15\%$. If the focus depth stretching technique works we can hope on further reduction of depolarization. The two stage system with the same parameters gives 100 times reduction of emittances (with the same restrictions). The maximum emittance at the entrance (the electron beam radius is two times smaller than the laser spot size) is about 10^{-3}cm·rad (the increase of this number is possible after some optimization).

For the cooling of the electron bunch train one laser pulse can be used many times. According to (7) $\Delta E/E = \Delta A/A$ and even 25% attenuation of laser power leads only to small additional energy spread.

The proposed scheme of laser cooling of electron beams seems very promising for future linear colliders and allows to reach ultimate luminosities. Especially it is useful for photon colliders, where collision effects allow considerable increase of the luminosity. Perhaps this method can be used for X-ray FEL based on high energy linear colliders.

ACKNOWLEDGEMENTS

I would like to thank Z.Parza, the organizer of the Program "New Ideas for Particle Accelerator" at ITP, UCSB, Santa Barbara, supported with National Science Foundation Grant NO PHY94-07194. I am grateful to A. Skrinsky for useful discussions and critical remark concerning polarization. Also I would like to thank D. Cline, S. Drell, I. Ginzburg, J.Irving, G. Kotkin, D. Leith, B.Palmer, A. Sessler, V. Serbo, B. Richter, P. Zenkevitch for useful discussions.

REFERENCES

1. Low et al., International Linear Collider Technical Review Committee Report, SLAC-Rep-471(1996)
2. I.Ginzburg, G.Kotkin, V.Serbo, V.Telnov,*Pizma ZhETF*, **34** (1981)514; *JETP Lett.* **34** (1982)491.
3. I.Ginzburg, G.Kotkin, V.Serbo, V.Telnov,*Nucl.Instr. & Meth.* **205** (1983) 47.
4. V.Telnov,*Nucl.Instr.&Meth.A* **294** (1990)72.
5. V.Telnov, *Nucl.Instr.&Meth.A* **355**(1995)3.
6. *Proc.of Workshop on $\gamma\gamma$ Colliders*, Berkeley CA, USA, 1994, *Nucl. Instr. &Meth. A* **355**(1995)1-194.
7. P.Chen,V.Telnov,*Phys.Rev.Letters*, **63** (1989)1796.
8. V.Telnov, *Proc.of Workshop 'Photon 95'*, Sheffield, UK, April 1995, p.369. '
9. H.Wiedemann, *Particle Acc. Physics: basic principles and linear beam dinamics*, Springer-Verlag, 1993.
10. D.Strickland and G.Mourou, *Opt commun.* **56**(1985)219.

11. C.Clayton, N.Kurnit, D.Meyerhofer, *Proc. of Workshop on Gamma – Gamma Colliders,* Berkeley CA, USA, 1994, *Nucl. Instr.&Meth.A* **355**(1995) 121.
12. C.Travier, *Nucl.Instr.&Meth.A* **340**(1994)26.
13. V.Berestetskii, E.Lifshitz and L.Pitaevskii, *Quantum Electrodynamic,* Pergamont press, Oxord, 1982.
14. G.Kotkin, S.Polityko, V.Serbo, Yadernaya Fizika, v.59 (1996)

Parametric X-ray Radiation as Source of Pulsed, Polarized, Monochromatic, Tunable X-ray Beam

Z. Parsa[†1] and A.V. Shchagin [‡2]

[†]*Brookhaven National Laboratory*
901A Physics Dept., Upton, NY 111973-5000, USA
[‡]*Kharkov Institute of Physics and Technology,*
Kharkov 310108, Ukraine

Abstract. The parametric X-ray radiation (PXR) is a new type of monochromatic, polarized, directed, tunable radiation arising in the vicinity of Bragg directions of a crystal when relativistic particles are passing through the crystal. For the last ten years the PXR properties have been intensively investigated experimentally at moderate incident electron current. In this paper we consider main properties of PXR as a new source of a powerful monochromatic X-ray beam, excited by short high-current electron bunch in a single crystal. It is shown, that using a PXR, one can obtain X-ray beam power of about MW/sr during a short period of time at moderate incident electron energy. Experimental setup for generating pulsed PXR beam is suggested, and possibilities for application of pulsed polarized monochromatic X-ray beams are discussed.

I INTRODUCTION

In recent years the interest for creation of new generation of powerful pulsed X-ray source is increasing. Considerations and investigations of different methods of X-ray production for this purpose are in progress. Especially of interest is creation of powerful source of polarized X-ray beam. In this paper we will present and analyze properties of Parametric X-ray Radiation (PXR) as source of powerful pulsed polarized monochromatic X-ray beam.

PXR is coherent type of radiation and it is due to the interaction of electromagnetic field of incident particle with electron subsystem of crystal. From the theory [2] it may be seen that the PXR angular distribution has a sharp conical maximum (reflection) with the gap in the middle, found around the Bragg

---

[1)] Supported in part by US Department of Energy Contract No. DE-AC02-76CH00016
[2)] Supported in part under the Grant of Ukrainian Fund of Basic Research

direction. The angular size of the PXR reflection is about or greater than $\frac{1}{\gamma}$, where γ is the relativistic factor of incident particle. Earlier experimental investigations have been performed at moderate electron beam current. Here we will consider the case when PXR is excited by high current short pulse of electron beam (bunched electron beam), e.g. as in [17].

The radiation generated by the charged particle moving through a medium with a periodically varying permittivity was first predicted by Fainberg and Khizhnyak [1] in 1957. Later, the radiation of charged particles moving through a crystal in the X-ray band was investigated by Ter-Mikaelian [2] and several versions of theoretical descriptions of PXR were proposed e.g., in Refs. [3–6]. The experimental studies of PXR properties were began in the late 80-ies e.g., in Refs. [7–12], continued in 90-ies e.g., in Refs. [13–15], and are reviewed e.g., in Ref. [16].

II FORMULATION

In this section we will present some of the formulae we have used in our calculations. By analogy with the ordinary diffraction of X-rays in a crystal, there are two schemes for PXR generation: one is the Laue geometry and the other is the Bragg geometry. In both cases, the beam of charged relativistic particles with velocity $\vec{V}$ passes through a crystal-radiator, and the narrow PXR beam (reflection) of a conical shape is generated around the Bragg direction with respect to the crystallographic planes. Most of the experiments were carried out in the Laue geometry. This geometry is shown in Fig. 1.

The PXR quanta energy may be calculated according to the formula from ref. [2]

$$E_{CR} = \hbar\omega_{CR} = \frac{c\hbar \left| \vec{g} \cdot \vec{V} \right|}{c - \sqrt{\varepsilon_0}\vec{V}\cdot\vec{\Omega}}, \qquad (1)$$

where $\hbar$ is the Plank constant divided by 2π, ε_0 is the constant part of the medium permittivity, $\vec{g}$ is the reciprocal lattice vector, $\vec{\Omega}$ is the unit vector in the direction of the detector, c is the light velocity.

The width at half maximum of the PXR spectral peak for a detector of small angular size is

$$\Delta E_{CR} \approx E_{CR}\frac{\Delta\theta}{tg(\theta/2)}, \qquad (2)$$

where $\Delta\theta$ is the polar angular resolution of the experiment. The $\Delta\theta$ value should include the detector size and the beam spot size on the target, and also the electron beam divergence in the target (including electron multiple scattering). Formula (2) was derived and confirmed experimentally in ref. [9].

The differential yield is described by the formula for the number of quanta, derived in ref. [9] from the kinematical theory [2]

$$Y_{\vec{g}} = \frac{dN}{nd\Omega} = \frac{e^2 \omega L \left|\chi_{\vec{g}}(\omega)\right|^2}{2\pi\hbar\varepsilon_0^3 V \left(\frac{c}{\sqrt{\varepsilon_0}} - \vec{V}\cdot\vec{\Omega}\right)} \cdot$$

$$\left| \frac{\left(\omega\sqrt{\varepsilon_0}/c\right)\vec{\Omega} \times \left[(\omega\varepsilon_0/c^2)\vec{V} + \vec{g}\right]}{\left[\left(\omega\sqrt{\varepsilon_0}/c\right)\vec{\Omega_\perp} - \vec{g_\perp}\right]^2 + \left(\frac{\omega}{V}\right)^2\left[(1/\gamma)^2 + \left(\frac{V}{c}\right)^2(1-\varepsilon_0)\right]} \right|^2, \quad (3)$$

where dN is the number of quanta with frequency $\omega = \omega_{CR}$ (see formula (1)) emitted in the solid angle $d\Omega$ as n particles with the charge e and relativistic factor γ pass through a crystal of a thickness L; $\chi_{\vec{g}}(\omega)$ is the Fourier component of the variable part of permittivity; $\vec{\Omega_\perp}$, $\vec{g_\perp}$ are the components of $\vec{\Omega}$, $\vec{g}$ perpendicular to $\vec{V}$.

The factor L with the attenuation of PXR in target taken into account should be used in the form [11]

$$L = T_e \left|\frac{\vec{t}\cdot\vec{\Omega}}{\vec{t}\cdot\vec{v}}\right| \left[1 - \exp\left(-\frac{T}{T_e\left|\vec{t}\cdot\vec{\Omega}\right|}\right)\right] \quad (4)$$

where $T_e = T_e(\omega)$ is the e-fold attenuation length of radiation with frequency ω; T is the target plate thickness; $\vec{t}$ is the unit vector perpendicular to the target plane; $\vec{v} = \frac{\vec{V}}{V}$.

Calculations by formulae (1), (2), (3), (4) have been compared to experimental data on differential PXR yield in refs. [9,11,12,16] and a good agreement of calculated and measured data was found.

The background conditions in PXR spectra were discussed in ref. [12]. where it was shown that the resulting PXR spectral peak/background ratio in the vicinity of Bragg direction may be about or more than 100 even at a moderate incident electron energy.

The linear polarization directions are oriented radially around the reflection center. If the detector is of the same size or greater than the PXR reflections, the average polarization should be generally close to zero. However, by collimating the corresponding differential part of the PXR reflection, the X-ray beam with an arbitrary direction of linear polarization may be chosen for applications. This was confirmed experimentally in ref. [8]. In particular, if the reciprocal lattice vector $\vec{g}$ is in the detection plane, the linear polarization direction is in the detection plane too (parallel polarization [16]).

So, PXR is perspective source of monochromatic, tunable, polarized X-ray beam with a low spectral background. The mentioned PXR properties have been confirmed at moderate electron beam current for PXR quanta energies 5 - 400 keV at electron beam energy 15 - 1200 MeV [16].

III PULSED PXR

Below we will consider PXR as source of powerful monochromatic polarized X-ray beam, generated in a crystal by short electron bunch, when the intensity of PXR beam will be high enough during a short pulse.

The suggested experimental layout is shown in Fig.1. The beam of accelerated bunched electrons (labeled as 1) passes in vacuum through a target (labeled as 2). A thin silicon single crystal plate is used as a target 2. The crystallographic plane ($\bar{2}20$) is parallel to the plate surface. The target is placed in a 3-axis vacuum goniometer and is prealigned so that the $<\bar{2}20>$ axis should be parallel to the velocity vector $\vec{V}$ of the (beam) electrons, and the crystallographic plane (111) should be perpendicular to the detection plane and parallel to the vector $\vec{V}$. At $\phi = 0$ the crystallographic plane (111) is parallel to the electron beam and the reciprocal lattice vector $\vec{g} = <111>$ lie in the detection plane. The PXR reflection (labeled as 3) from planes (111) has an asymmetrical conical form and the Bragg direction (labeled as 4) is the cone axis. To obtain polarized X-ray beam we can use the part of PXR reflection, where PXR yield has maximum value. This part has conical form with full cone angle $1/\gamma$ (at moderate incident electron energy). The cone axis is at angle $\theta = \theta_B + 1/\gamma$. The angle of crystal rotation is $\phi = \theta_B/2$. The direction of linear polarization lies in the plane of the figure. Detector or consumer (labeled as 5) of pulsed polarized X-ray beam is shown in Fig. 1.

Target thickness is chosen to provide electron beam divergency in the target no more than $1/\gamma$. Target thickness $T = T_\gamma$ can be found from the equation

$$\frac{1}{\gamma} = \theta_s(T) \qquad (5)$$

where θ_s is the electron multiple scattering angle [2] in silicon. Note, that T_γ is independent of incident electron energy and is a function only of target material properties. For silicon single crystal $T_\gamma \approx 60 mkm$. We will use this thickness in our calculations.

To calculate power of pulsed PXR beam, we used a typical (for an available) parameter set for a bunched electron beam: charge of electron bunch $1nC$ and bunch duration of $333fs$ at electron beam energies 50, 100 and 200 MeV.

Target temperature is increasing when the target is working as crystal-radiator in electron beam. It has been shown in ref. [14], that Si single crystal can work as PXR radiator at average electron beam current of about 0.5 A at electron beam radius of 1 mm. In our case, average bunched electron beam current is function of accelerator repetition rate and repetition rate of real accelerators usually is much less than $0.5A/(1.6 \cdot 10^{-9}C) \sim 3 \cdot 10^8 \frac{bunches}{s}$. Therefore we do not need to take into account crystal-radiator heating and we use the room temperature for Si single crystal in our estimations.

In calculations of bunched PXR beam properties we used Eqs. (1), (3) from refs. [2,9], which was confirmed experimentally in previous investigations

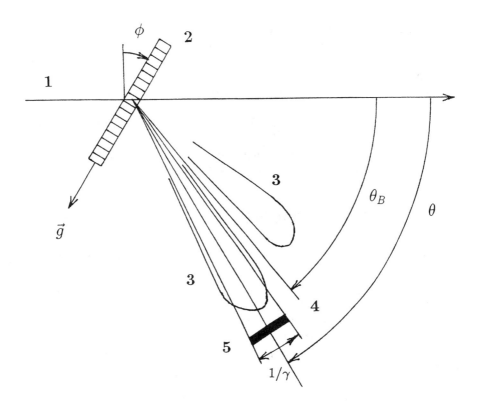

Fig. 1

Fig. 1. The experimental setup for generation of powerful, pulsed, monochromatic, polarized x-ray beam. Electron beam 1 is passing through a crystal slab 2. Vector $\vec{g} = <111>$ is the reciprocal lattice vector for the crystallographic planes (111) shown by hatching on Si single crystal target 2. Target 2 is placed in a vacuum 3-axis goniometer and the angle ϕ is the angle of crystal rotation. The PXR reflection 3 has asymmetrical conical form around of Bragg direction 4. The X-ray beam polarized in the plane of figure is going at angle $\theta_B + \frac{1}{\gamma}$ to consumer 5 in full cone angle equal to $1/\gamma$.

[9,11,12,16] at moderate electron beam current. Results of the calculations of polarized X-ray beam power in units MW/steradian are shown in Fig.2, as a function of required energy of X-ray quanta. These results may be used to estimate the power flux in term of full cone angle of $\leq 1/\gamma$. For every chosen quanta energy it is necessary to provide a required setup geometry (angles ϕ, θ, see Fig. 1). For the data, represented in Fig. 2, angles θ are in the range 40 - 170 mrad. The width of PXR spectral line may be easily calculated using formula (2). It is seen from Fig. 2, that intensity of polarized PXR beam is high enough and it may be applied in different fields. Pulsed PXR with short pulse, high intensity electron beam are discussed in refs. [17-19].

IV APPLICATIONS OF PULSED PXR BEAM.

During the last decades numerous applications of a monochromatic X-ray beam have been developed. Now the most popular source of an X-ray beam is synchrotron radiation. But synchrotron radiation usually is not able to provide high power X-ray pulses. As can be seen from Fig.2, PXR is a perspective source of powerful short pulses of polarized, monochromatic tunable X-ray beam. Duration of X-ray pulses is determined by duration of electron beam bunch. Power of the X-ray pulse is proportional to the charge of electron bunch. Direction of linear polarization of X-ray beam at consumer 5 is in the plane of Fig. 2.

A powerful polarized pulsed X-ray beam, suggested in this paper, may be applied in different areas of science and technology. Following are some of the applications, where the polarized X-ray beam may be used:

1. For measurements of bunched electron beam parameters
2. For measurements of short-time crystal-radiator properties
3. For medical and biological applications and diagnostics
4. For investigations of femtosecond processes in solid state.
5. For precision calibration of spectral, angular and polarization properties of new generation of astrophysical space X-ray and gamma-ray telescopes and other X-ray equipments.
6. For pumping of X-ray laser. Femtosecond pulses are especially convenient for this purpose, because pulse duration is comparable with lifetime of K- and/or L-shell vacancy in atoms.
7. For control and detecting of nuclear materials in cosmic space, atmosphere, airports, railways and so on. Pulsed PXR monochromatic X-ray beam is very convenient instrument for control of nuclear materials circulation and proliferation in the world. For example, linac-based PXR source of X-ray beam may be launched to cosmic space for remote detecting of nuclear materials in satellites. Scanning of X-ray beam energy around the absorption K-edge may be applied for detection of nuclear materials.

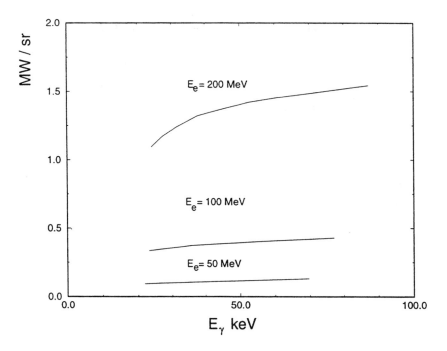

Fig. 2

Fig. 2. Power [MW/sr] of polarized monochromatic X-ray beam as function of photon energy for incident electron beam energy of 50, 100, 200 MeV. Duration of X-ray beam pulse is 333 fs. Charge of electron bunch is 1 nC. Angle θ of X-ray beam generation is in the range 40-170 mrad. Thickness of Si single crystal is 60 mkm.

8. For remote investigations of chemical composition of asteroids and moon, using monochromatic X-ray beam for excitation of characteristic X-rays. Linac-based PXR source of X-ray beam now is the only single perspective source of monochromatic X-ray beam in space. (Synchrotron may not be launched now due to its geometry and size.)

Development and investigations of PXR as a source of powerful polarized X-ray beam will be useful for the above mentioned and other applications. Real crystals have sets of numerous crystallographic planes. As a result, a set of numerous powerful polarized X-ray beams with different energies will be propagated from a crystal, situated in the electron beam, in different directions. Many of those X-ray beams may be utilized simultaneously much like the synchrotron radiations is utilized from different parts of a storage ring. Short X-ray pulses are very convenient for improvement of measurement sensitivities at remote locations. Linear polarization of X-ray beam may be utilized for the same purpose.

We (Z.P.) thank A. Shchagin and ask interested readers to direct their questions re data, chronology of the experiments and papers in the Former Soviet Union to him. We acknowledge and thank D. Rule for discussions and ITP Program "New Ideas for Particle Accelerators" for providing partial support for this work - with the National Science Foundation under Grant No. PHY94-07194.

REFERENCES

1. Fainberg Ya. B. and Khizhnyak N.A., Zh. Ehksp. Teor. Fiz. **32**, 883-895 (1957).
2. Ter-Mikaelian M.L., High-Energy Electromagnetic Processes in Condensed Media (Wiley-Interscience, New York, 1972).
3. Baryshevskii V.G., Channeling, Radiation and Reactions in Crystals at High Energies (BGU publ., Minsk, 1982) [in Russian].
4. Garibian G.M. and Yang C., X-ray Transition Radiation (Acad. Sciences of Arm. SSR, Yerevan, 1983) [in Russian].
5. Dialetis D., Phys. Rev. A **17**, 1113-1122 (1978).
6. Feranchuk I.D. and Ivashin A.V., J. Physique **46**, 1981-1986 (1985).
7. Adishchev Yu.N., Didenko A.N., Mun V.V., Pleshkov G.A., Potylitsin A.P., Tomchakov V.K., Uglov S.R. and Vorobiev S.A., Nucl. Instr. and Meth. B **21**, 49-55 (1987).
8. Adishchev Yu.N., Verpinov V.A., Potylitsin A.P., Uglov S.R. and Vorobiev S.A., Nucl. Instr. and Meth. B **44**, 130-136 (1989).
9. Shchagin A.V., Pristupa V.I. and Khizhnyak N.A., Phys. Lett. A **148**, 485-488 (1990).
10. Morokhovskii V.L. and Shchagin A.V., Zh. Tekh. Fiz. **60**, no. 5, 147-150 (1990); In English: Sov. Phys. Tech. Phys. **35**, 623-624 (1990).

11. Shchagin A.V., Pristupa V.I. and Khizhnyak N.A., Proceedings of International Symposium on Radiation of Relativistic Electrons in Periodical Structures, September 6-10, 1993, Tomsk, Russia, p.62-75. See also [16].
12. Shchagin A.V., Pristupa V.I. and Khizhnyak N.A., Nucl. Instr. and Meth. B **99**, 277-280 (1995).
13. Fiorito R.B., Rule D.W., Maruyama X.K., DiNova K.L., Evertson S.J., Osborne M.J., Snyder D., Rietdyk H., Piestrup M.A. and Ho A.H., Phys. Rev. Lett. **71**, 704-707 (1993).
14. Fiorito R.B., Rule D.W., Piestrup M.A., Qiang Li, Ho A.H. and Maruyama X.K, Nucl. Instr. and Meth. B **79**, 758-761 (1993).
15. Fiorito R.B., Rule D.W., Piestrup M.A., Maruyama X.K., Silzer R.M., Skopik D.M. and Shchagin A.V., Phys. Rev. E **51**, Part 1, R2759-R2762 (1995).
16. Reviews about recent investigations of PXR properties: A.V. Shchagin, The Future of Accelerator Physics, The Tamura Symposium Proceedings, Austin TX, November 1994, AIP Press, pp.359 - 377; A.V. Shchagin, N.A. Khizhnyak, will be published in October 1996 issue in NIM B; A.V. Shchagin, X.K. Maruyama, Parametric X-rays. Will be published as a chapter in book: Accelerator based atomic physics technique and applications, edited by S.M. Shafroth and J.C. Austin, AIP Press, 1997.
17. Z. Parsa, "Pulsed Parametric X-Radiation", Presentation UCSB/ITP96 Santa Barbara, Report (1996).
18. Z. Parsa, "New Ideas for Particle Accelerators" Program Summary Report, Santa Barbara, Ca. (1997).
19. Z. Parsa, "Pulsed Parametric X-Radiation (PPXR) – A New Alternative to Synchrotron Source", American Physical Society Presentation, APS-APR97 LogNo. 6023, (April 1997).

Advanced Linacs for a Linac Ring Collider

David B. Cline

Center for Advanced Accelerators
Department of Physics and Astronomy, Box 951547
University of California, Los Angeles, Los Angeles, CA 90095-1547 USA

Abstract. We describe the concept of a high-luminosity asymmetric ϕ factory and show how high-intensity, strangeness-tagged $K^0/\overline{K}^0$ particles can be obtained for CPT tests. We also describe the concept of an asymmetric ϕ factory using linac storage rings. Finally, we describe the study of a powerful linac storage-ring collider that could give an asymmetric-ϕ-factory luminosity of 10^{34} cm^{-2} s^{-1}, making for the ultimate CPT tests. We describe the concept of a high-luminosity asymmetric ϕ factory and show how high-intensity, strangeness-tagged $K^0/\overline{K}^0$ particles can be obtained for CPT tests. We also describe the concept of an asymmetric ϕ factory using linac storage rings. Finally, we describe the study of a powerful linac storage-ring collider that could give an asymmetric-ϕ-factory luminosity of 10^{34} cm^{-2} s^{-1}, making for the ultimate CPT tests.

INTRODUCTION

The final frontier of particle physics may occur at the Planck scale (~ 10^{19} GeV). In as much as there is no known method to produce such energies, two tests have been proposed to study this region:

1. If proton decay is observed, and if it proceeds through the exchange of an X particle of mass of ~ 10^{16} GeV, there could be interference effects due to the Planck mass exchange.

2. CPT could be violated at the Planck scale if locality of the fields is broken (1).

Both of these types of measurements are extremely hard to do. We discuss here the search for CPT violation using a special type of ϕ factory called an "asymmetric ϕ factory" (2,3).

In Fig. 1, we trace the road to ϕ factories (this is from work we did in the late 1980s) (1). Note that it has taken nearly 30 years to actually build a ϕ factory (DaϕNE), which should operate in 1997. Unfortunately, the UCLA symmetric ϕ factory (which used an entirely different approach) was not approved by the US Department of Energy in 1991.

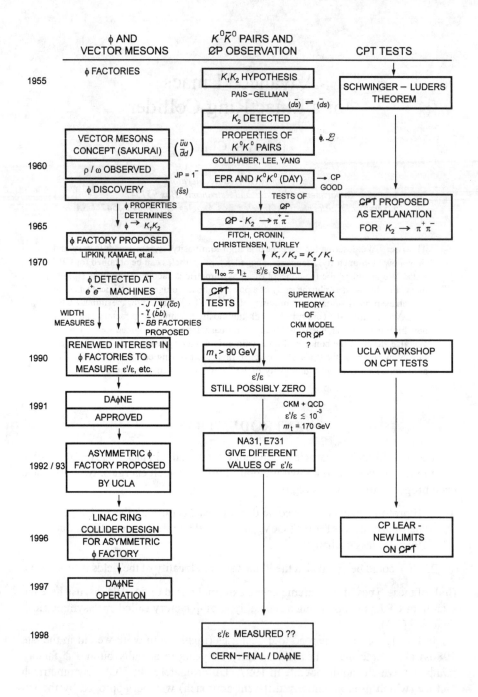

FIGURE 1. Road to φ factories and CPT tests.

The test for CPT violation that uses $K^0/\overline{K}^0$ beams attempts to measure the CPT violating parameters shown in Fig. 2. Since there is no real theory of CPT violation, it is hard to "predict" the nature of a signal for this violation. However, we know that present limits are not all that far from the possible Planck-scale violation effects.

SYMMETRIC VS ASYMMETRIC φ FACTORIES

Many virtues of a symmetric φ factory were emphasized in a 1990 workshop at UCLA and published in the proceedings (1). The concept of an asymmetric φ factory arose in 1992 at another workshop at UCLA (2); there was also a workshop on symmetric φ factories in 1993 at UCLA. Figure 3 gives a comparison of the K_s^0 decay length for the two (symmetric and asymmetric φ factories), showing the large difference. This is the key parameter to be able to tag $K^0/\overline{K}^0$ in a model-independent way.

A second useful aspect of an asymmetric φ factory is the possibility of tagging the strangeness of the beam by the strong interaction of the K^0 or $\overline{K}^0$, e.g.,

$$\phi \to K^0/\overline{K}^0 \quad , \quad \overline{K}^0 + p \to n + \overline{N}^+ \quad ,$$

$$\phi \to K^0/\overline{K}^0 \quad , \quad K^0 + p \to K^+ + n \quad .$$

If the $K^0/\overline{K}^0$ is tagged at $t = t_a$, the other particle in the φ decay must be the antiparticle at that time, thus giving a particle/antiparticle-identified beam (3). A detailed comparison of tagged $K^0/\overline{K}^0$ beams leads to extremely precise CPT tests, some of which are listed in Table 1. We emphasize that this tag does not rely on the use of the $\Delta S = \Delta Q$ rule.

A third feature of an asymmetric φ factory is the possible realization of very high luminosity in a linac storage-ring collider, which we will now discuss.

Table 2 lists the various types of asymmetric φ factories with some pros and cons for the different schemes (2). In this paper, we only discuss the linac collider option with a high-energy e^+-beam storage ring and a low-energy e^--beam.

The basic idea to enhance the luminosity of a linac ring collider is shown in Fig. 4(A) and (B). A high-current, high-energy stored e^+ beam interacts with a lower-current, lower-energy e^- beam and, in the process, the e^- beam is pinched into the e^+ beam, giving a high-density e^+e^- beam, with the resulting e^- beam fully disrupted, as shown in Fig. 4(B). This process occurs at high frequency, giving rise to a possible high luminosity. Care must be taken that the low-current e^- beam does not cause other types of damage (e.g., a displacement, etc.) to the e^+ beam. We believe these and many other problems can be overcome if the high-energy e^+ storage ring has a very fast damping time to erase such displacements and other problems. Other schemes for asymmetric φ factories have been proposed (4). A key issue is the dynamics of the beam–beam interaction, which has been studied elsewhere (5,6).

$$|K_s^0> \frac{1}{\sqrt{2(1+|\epsilon+\delta|^2)}}\{[1+(\epsilon+\delta)]|K^0> + [1-(\epsilon+\delta)]|\overline{K}^0>\} \ ,$$

$$|K_L^0> \frac{1}{\sqrt{2(1-|\epsilon+\delta|^2)}}\{[1+(\epsilon-\delta)]|K^0> - [1-(\epsilon-\delta)]|\overline{K}^0>\} \ ,$$

$$\eta_\pm = \epsilon_0 + \epsilon' \ , \qquad \epsilon_0 = \epsilon - \delta - \lambda_0 \ , \qquad \lambda_0 = \frac{(\overline{A}_0 - A_0)}{(\overline{A}_0 + A_0)} \ ,$$

where δ = CPT in mass matrix, and λ_0 = CPT in direct amplitude.

FIGURE 2. K_s^0/K_L^0 formulation with CPT violating parameters indicated.

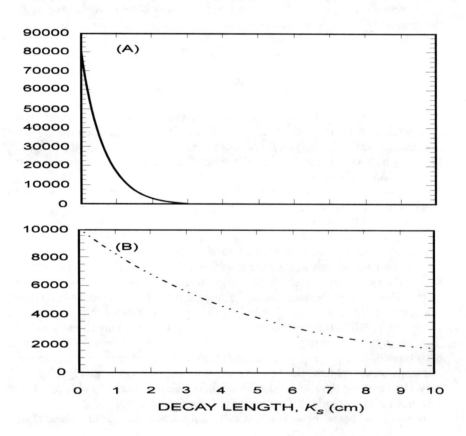

FIGURE 3. K_s decay length for (A) symmetric and (B) asymmetric ϕ factories.

TABLE 1. Methods to Tag the K^0 Strangeness with an Asymmetric ϕ Factory

If $\phi \gg \overline{K}_i^0 K_j^0$ ($t_i = t_j$),

(i) $\overline{K}_{i,j}^0 + p \rightarrow \Lambda + \pi^+$ For flavor ID in a connector
(do not need $\Delta S = \Delta Q$ rule for test)

(ii) $\overline{K}_i^0 \rightarrow \pi^+ e^- \overline{v}$, $\Delta S = \Delta Q$ rule assumed $t_i = t_j$
$\overline{K}_j^0 + p \rightarrow \Lambda + \pi^+$ Flavor ID

(iii) $\pi^+ e^- \overline{v}_e$,
$\pi^- e^+ v_e$ $\Delta S = \Delta Q$ rule assumed $t_i = t_j$; unique e^- ID using forward C counters and magnetic spectrometer

(iv) $\phi \gg K_i K_j$ Symmetric or asymmetric ϕ factory
$\downarrow \downarrow$
$\pi^+\pi^-$ ($t_i = t_j$); possible serious background from $\phi \rightarrow K_S K_S \gamma$

TABLE 2. Types of Asymmetric ϕ Factories

Type	Advantages	Disadvantages
1. Low-energy e^+ storage ring or accumulator and high-energy e^- linac	(A) Rapid damping time e^+ means reduced instabilities (B) Easy to produce low-energy e^+ storage ring	(A) Need high rep rate e^- linac (B) High-energy linac is expensive
2. Low-energy e^- linac and high-energy e^+ storage ring	(A) Low-energy superconducting linac (B) e^- trapped in e^+ bunch	(A) Expensive e^+ source (B) Damping time of e^+ ring may allow buildup of instabilities
3. e^- linac or e^+ linac	Requires novel e^+ source for e^+ linac	
4. e^- storage ring or e^+ storage ring	More difficult than a symmetric e^+e^- collider	

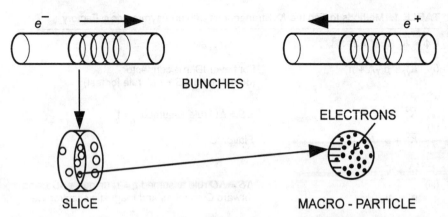

FIGURE 4. (A) Collision model where colliding bunches are divided into slices, with each slice populated with a random distribution of macroparticles containing the charged particles. The overall behavior of the particles in the bunches is approximated by the behavior of the macroparticles.

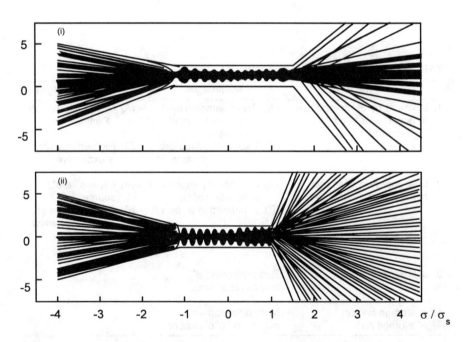

FIGURE 4. (B) Electron trajectories at the collision point (2σ contour) viewed in the test frame of the positron bunch with (i) Gaussian and (ii) parabolic longitudinal distributions.

A POSSIBLE, POWERFUL LINAC STORAGE-RING ASYMMETRIC φ FACTORY

The linac ring scheme that is being studied at UCLA is shown in Fig. 5. A powerful compact storage ring that uses 7-T superconducting bending magnets is used for the positron beam. Table 3 gives some parameters for the storage ring. Note that the damping times are 200–400 μs! An advanced linac such as Tesla, which is shown schematically in Fig. 5 (6), is used for the e^- beam. Table 4 gives the required parameters of the linac and the overall parameters required to reach a luminosity of 10^{34} cm^{-2}s^{-1}.

TESTS OF QUANTUM MECHANICS

The asymmetric φ factory concept provides for novel tests of quantum mechanics. One possibility, which was devised by P. Eberhard, is shown in Fig. 6 (7). The rates in the various configurations should show clearly the effects of quantum interference, which has never been measured for a $K^0/\overline{K}^0$ correlated pair (7).

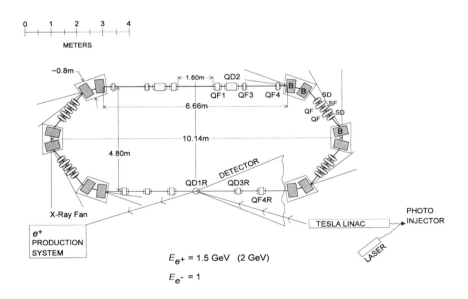

FIGURE 5. UCLA ultra-compact light source and asymmetric φ factory.

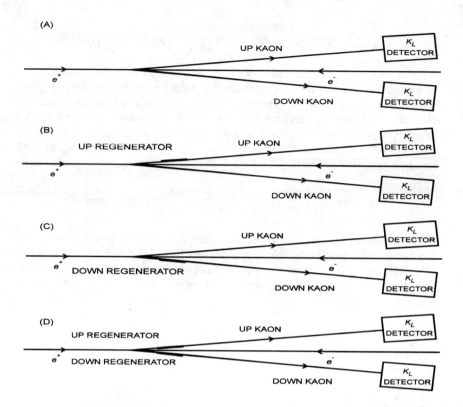

FIGURE 6. Setups for the destructive interference test.

TABLE 3. Lattice Parameters for the Compact Storage Ring

Maximum stored-beam energy (GeV)	1.5
Projected injection-beam energy (GeV)	0.1
Projected beam current (mA)	-200
Circumference (m)	26
Bend radius (m)	0.7257
Dipole bend angle (deg)	30
Integrated dipole induction* (T m)	2.6197
Dipole central induction (T)	6.894
Dipole magnetic length along the bend (m)	0.38
Critical energy (keV)	10.3
Horizontal natural emittance* (µm)	2.34
Vertical coupled emittance* (µm)	1.17
Vertical operating emittance* (µm)	0.0234
Horizontal tune	3.17
Vertical tune	2.57
Horizontal chromaticity	-2.22
Vertical chromaticity	-5.24
Maximum horizontal beta function (m)	3.09
Maximum vertical beta function (m)	6.66
Maximum dispersion (m)	1.29
Energy loss per turn* (MeV)	0.617
RF voltage (MV)	2.5
RF frequency (MHz)	499
Energy spread (parts in 1000)	1.52
Bunch length rms* (mm)	30
Horizontal damping time* (ms)	0.412
Vertical camping time* (ms)	0.422
Energy damping time* (ms)	0.213
Quantum lifetime (s)	2.2×10^8

*At the maximum beam-energy design.

Table 4. Asymmetric φ-Factory Linac Calculations

No. of electrons	$N_e = 10^{10}$
Electron avg repetition rate	$f_e = 5 \times 10^5$
No. of positrons	$N_p = 10^{11}$
Positron period	$T_p = 87 \times 10^{-9}$
Positron avg repetition rate*	$f_p = T_p^{-1}$
Electron current	$I_e = 8 \times 10^{-4}$ ($I_e = N_e \times 1.6 \times 10^{-19} \times f_e$)
Duty cycle of electrons	$\eta = 0.044$ ($\eta = f_e/f_p$, 4 × Tesla test facility)
Accelerating gradient in Tesla linac	$E_{acc} = 1.5 \times 10^7$
Electron beam energy (eV)	$E_e = 1.8 \times 10^8$
Active length of linac	$L_{acc} = 12$ m ($L_{acc} = E_e/E_{acc}$)
Normalized emittance of electrons	$\epsilon_e = 2 \times 10^{-6}$
Physical electron emittance	$\epsilon = 5.678 \times 10^{-9}$ ($\epsilon = \epsilon_e/\gamma$, where $\gamma = E_e/5.11 \times 10^5$)
Positron energy (eV)	$E_p = 1.5 \times 10^9$
Luminosity (cm^{-2} s^{-1})	$\mathcal{L} = 10^{34}$

*Same as electron avg repetition rate during linac pulse.

REFERENCES

1. *Testing CPT and Studying CP Violation at a φ Factory* (Proc. of UCLA Workshop, Los Angeles, April 1990), D. Cline (ed.), *Nucl. Phys. B*, **24A** (1991).
2. *Gamma Ray-Neutrino Cosmology and Planck Scale Physics*, D. Cline (ed.), Singapore: World Scientific, 1993, Chap. VI (proc. of a mini-workshop on asymmetric φ factories) pp. 303–346.
3. Cline, D., "A New Concept for an Asymmetric φ Factory to Test CPT and Study K_s^0 Mesons," in *Proceedings of the XXVI International Conference on High Energy Factories to Test CPT and Quantum Mechanics*, Singapore: World Scientific Press, 1994.
4. Eberhard, P., and Chattopadhyay, S., "Asymmetric φ Factories-A Proposed Experiment and Its Technical Feasibilty," in *Proceedings*, 15th Int. Conf on High Energy Accelerators (Hamburg, Germany, July 1992).
5. Grosse-Wiesmann, P., *Nucl. Instrum. & Meth. A*, **274**, 21 (1989).
6. Johnson, C. D., "The Incoherent Beam-Beam Effect in Linear-On-Ring Colliders," in *Proceedings* (IEEE Particle Accelerator Conference, San Francisco, May 1991). (This work was carried out with A. Garron, M. Green, and J. Rosenweig).
7. Eberhard, P., "Testing the Non-Locality of Quantum Theory in Two K^0 System, Lawrence Berkeley Lab. report 32100, 1992.

High Current Short Pulse Ion Sources[1]

Ka-Ngo Leung

Lawrence Berkeley National Laboratory
University of California
Berkeley, California 94720

Abstract. High current short pulse ion beams can be generated by using a multicusp source. This is accomplished by switching the arc or the RF induction discharge on and off. An alternative approach is to maintain a continuous plasma discharge and extraction voltage but control the plasma flow into the extraction aperture by a combination of magnetic and electric fields. Short beam pulses can be obtained by using a fast electronic switch and a dc bias power supply. It is also demonstrated that very short beam pulses (~ 10 μs) with high repetition rate can be formed by a laser-driven LaB_6 or barium photo-cathode.

INTRODUCTION

High current short pulse ion beams are required in many accelerator facilities, such as Fermi Laboratory, LANSCE at Los Alamos, KEK in Japan, DESY at Hamburg, the SSC and the Proton Therapy Machine at Loma Linda Hospital. Future Spallation Neutron Sources (the NSNS, the ESS and the LANSCE upgrade) as well as compact neutron tubes will be operating with even higher current intensities and duty factor. In order to meet these future requirements, new ion sources or source operating schemes are being developed in some of the accelerator laboratories.

Ion beams can be conveniently extracted from a plasma generator. Short beam pulses are normally formed by modulating the arc discharge with an electronic switch. Figure 1 shows the experimental setup when a multicusp ion source is employed. The multicusp generator is capable of generating large volumes of uniform, high density and quiescent plasmas. It was originally developed to produce multi-amperes of deuterium ions with pulse length as long as 30 sec for neutral beam heating of tokamak fusion plasmas.

---

[1] Work supported by U.S. Department of Energy under Contract No. DE-AC03-76SF00098.

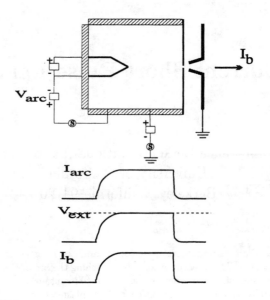

FIGURE 1. Pulsed mode operation for the multicusp ion source.

During the last decade, the multicusp source has been modified to produce high current, short-pulse ion beams for particle accelerator applications. In particular, the filament cathode in the source chamber is now replaced by an antenna coil and the plasma is produced by a 2 MHz RF induction discharge. It has been demonstrated that the RF driven multicusp source can be operated in pulsed mode to provide high intensities of either positive or negative ion beams.

Ion beams formed by RF discharge are typically limited to pulse lengths greater than 20 μs. Several new schemes are now being tested to generate ion beams with shorter pulses. This paper describes two techniques for short pulse ion beam formation. The first scheme employs a fast electronic switch together with a relatively low bias voltage on the plasma electrode. The second approach utilizes a laser-driven photocathode as a source of primary ionizing electrons.

EXPERIMENTAL SETUP

Generation of a RF discharge by placing the induction coil (or antenna) inside a multicusp source chamber was tested at Berkeley and Garching for neutral beam applications. [1] In 1991, a new RF driven H$^-$ source was developed at the Lawrence Berkeley National Laboratory (LBNL) for use in the injector unit of the SSC. [2] The source chamber is a thin walled (4-mm-thick)

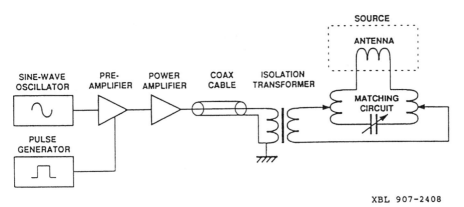

FIGURE 2. Schematic diagram of the complete RF power system.

copper cylinder (10-cm-diam by 10-cm-long) surrounded by 20 columns of samarium-cobalt magnets which form a longitudinal line-cusp configuration for plasma confinement. The magnets in turn are enclosed by a anodized aluminum cylinder, with the cooling water circulating around the source between the magnets and the inner housing wall.

A pair of permanent magnet filter rods can be installed near the first or plasma electrode to enhance the production of atomic H^+ or N^+ ions and volume-produced H^- ions. The black flange has four rows of magnets cooled by drilled water passages and contains all the required feedthroughs and ports, including a gas inlet, antenna feedthrough, and a 1-cm-diam opening for a quartz rod serving as a light pipe or window.

The source plasma is generated inductively by a two turn porcelain-coated copper coil antenna. The porcelain coating can survive months of operation without any significant deterioration. This two-turn induction coil is connected in series to a matching network as shown in Fig. 2. The 2 MHz RF signal, generated by a digital synthesizer is sent to a preamplifier, and then to the main power amplifier. The RF power output of the main amplifier can be controlled by changing the amplitude and frequency of the synthesizer signal. Maximum efficiency is achieved when the output voltage and current of the main RF amplifier are in phase and operating at a 50 Ω impedance. The RF power travels through a 50 Ω coaxial cable to the isolation transformer and matching network, and antenna. The matching network matches the impedance of the antenna coil immersed in the plasma to the 50 Ω impedance of the amplifier.

When the source is operated in pulsed mode, a small tungsten filament can be used to generate some electrons to aid in plasma ignition. However, the filament has a limited lifetime and contributes tungsten impurities to the plasma. It has been demonstrated that the ultraviolet light from a nitrogen laser impinging upon a magnesium target can provide enough photo emission electron to ignite the plasma. [3] Recently, it is shown that the more expensive laser could be abandoned in favor of an inexpensive xenon flash lamp. [4] The timing of the flash with respect to the RF pulse affects the plasma starting. In normal operation, the RF amplifier and the flash lamp are triggered at the same time. This will enable the plasma to be formed at the very beginning of the pulse and thus avoid the porcelain coating to be damaged by high RF antenna voltages.

EXPERIMENTAL RESULTS

Pulsed H⁻ Ion Beam Operation

A multicusp source has been operated with a RF induction discharge to generate volume-produced H⁻ ions. The SSC RF-driven source routinely provided 35 kV, > 35 mA H⁻ beams with a normalized rms emittance less than 0.1 π-mm-mrad. The source was typically operated with 100 μs beam pulse width at a 10 Hz repetition rate. The H⁻ output current of a RF-driven multicusp ion source can be increased by introducing a trace amount of cesium into the collar region. Most recently, the SSC RF-driven H⁻ source was modified to enhance the H⁻ output for testing a high current LINAC. A collar with eight cesium dispensers was installed at the exit aperture. A plasma grid heater element controls the temperature of the cesiated surfaces and the rate of cesium dispensation. With this arrangement, beam current in excess of 100 mA and e/H⁻ ratios close to one have been observed (Fig. 3). [5]

The extraction of H⁻ ions is accompanied by a large amount of electrons, which must be removed from the beam before further acceleration. The remaining H⁻ ion beam must be focused properly to match the acceptance of the next element of the accelerator system, e.g., a radio-frequency-quadrupole (RFQ). Recently, a permanent-magnet insert structure for the removal of electrons from pulsed, extracted negative ion beams has been developed at LBNL. [6] The simulated output of Fig. 4 shows that the electrons are removed from the H⁻ beam and made to impinge on the second electrode. [7] The trajectories of the H⁻ ions are not noticeably perturbed.

Source Operation with Inert and Diatomic Gases

The RF multicusp source has been tested with inert gas plasma such as He, Ne, Ar. Kr, and Xe. Figure 5 shows the extractable positive ion current density as a function of RF power. The optimum source pressure is typically below 1 mTorr. It can be seen that the output currents increase linearly with RF input power. In most cases, the extractable ion current density can be as high as 1 mA/cm^2 at approximately 50 kW of RF input power.

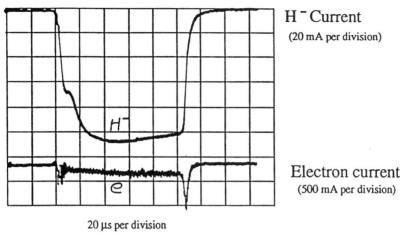

FIGURE 3. Oscilloscope traces showing a H- current of 100 mA.

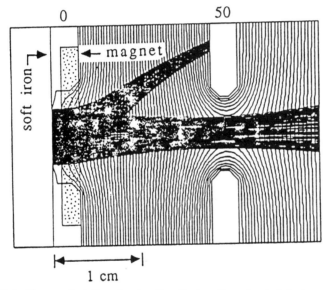

FIGURE 4. Planar calculation showing the effect on the extracted electrons due to a pair of permanent magnets located inside the first electrode.

The hydrogen ion species composition in the RF driven source has also been investigated. With a magnetic filter in place, the H^+ ion concentration is greater than 97% for an RF input power of 30 kW. The highest current density achieved is about 1.5 mA/cm^2.

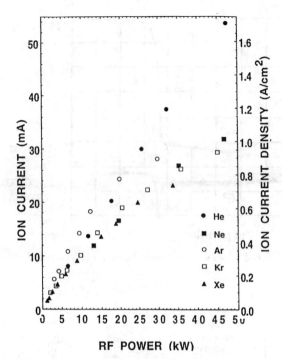

FIGURE 5. Extracted beam current and density as a function of RF power for various inert gas plasmas.

For a nitrogen discharge, the atomic N^+ ion concentration also increases with the RF input power. A nearly pure (> 98%) N^+ ion beam with current densities in excess of 500 mA/cm^2 has been obtained when the magnetic filter is employed. Similar results are obtained when other diatomic gases (such as oxygen) are used for the discharge.

Compact High Intensity Neutron Generator

A 25-mm-diameter RF-driven ion source has been developed for a high output, compact neutron generator. [8] This generator is designed to provide a 14 MeV neutron flux of 10^9 n/s, utilizing the D-T fusion reaction. Due to thermal and power constraints, the ion source is operated in pulse mode. This source is operated at 10 - 20 μs pulse width (Fig. 6) and repetition rate of 100 Hz. It is capable of producing the necessary extractable hydrogen current density of 800 mA/cm^2 with a monatomic species yield over 94% at a source pressure as low as 5 mTorr. [9] Thus, the new ion source demonstrates a substantial improvement over the Penning ion source used in the conventional sealed-tube neutron generators.

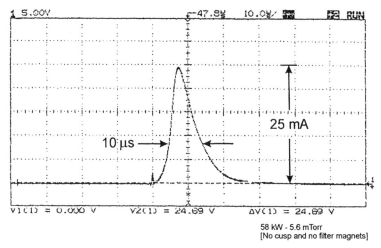

FIGURE 6. Beam pulse shape obtained from a subcompact RF driven ion source.

Pulsed Beam for Induction Linacs

In order to minimize the required volt-seconds, induction linacs are operated with short beam pulses of about 3 μs in width. Switching on ion beams by pulsing the discharge is too slow due to plasma formation time. An alternative approach is to keep the discharge and the extraction voltage at steady state while the plasma flow into the extraction aperture is controlled by a combination of magnetic and electric fields. Short beam pulses with high repetition rate can be generated with a combined arrangement of fast electronic switches and a dc bias power supply.

The dc power supply and the electronic switch are installed between the plasma electrode and the source chamber wall. In normal operation, the plasma electrode is connected to the anode walls. If a positive voltage ($\sim$ 100 V) is suddenly applied to the plasma electrode, the plasma in front of the exit aperture is pushed away by the electrostatic field and the positive ion current disappears (Fig. 7) . The time response of the beam intensity follows closely the applied voltage. A voltage modulator with rise/fall time less than 100 ns is now being developed at LBNL to reduce the beam pulse width.

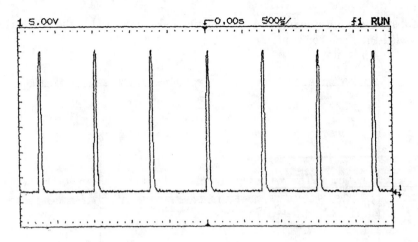

FIGURE 7. Micro beam pulses obtained by pulsed biasing the plasma electrode. Each beam pulse is about 40 μs in width.

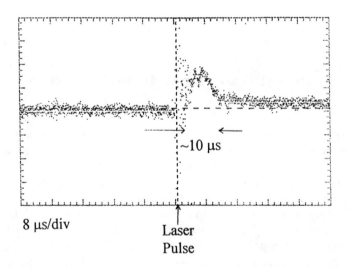

FIGURE 8. Langmuir probe signal of a pulsed plasma generated by a laser-driven photocathode.

Formation of Short Beam Pulses by Laser Driven Photocathodes

The feasibility of laser-induced photo-electron emission as a driving mechanism for short plasma pulse production has recently been investigated at LBNL with an excimer laser. [10] The exciting laserpulse energy ranged from 300 mJ to 500 mJ with a photon energy of 5 eV and a pulse width of approximately 40 ns. Two low work function materials, LaB_6 and barium, were used as cathode materials. A significant increase in the photo-emitted electron current was obtained when a low density plasma was present; the highest observed photo-emission current exceeded 140 A. Preliminary results (Fig. 8) showed that very short ion current pulses (~ 10 μs) can be formed by this laser-driven photo-emission scheme.

ACKNOWLEDGMENTS

The author would like to thank Y. Lee, L. T. Perkins, and D. S. Pickard for their assistance in preparing this manuscript. This work is supported by U.S. Department of Energy under Contract No. DE-AC03-76SF00098.

REFERENCES

1. M. C. Vella, K. W. Ehlers, D. Kippenhan, P. A. Pincosy, R. V. Pyle, W. F. DiVergilio, and V. V. Fosnight, J. Vac. Sci. Technol. A3, 1218 (1985). J. H. Feist et al., 14th Symposium on Fusion Technology, Avignon 1986 (unpublished), p. 1127.
2. K. N. Leung, G. J. DeVries, W. F. Divergilio, R. W. Hamm, C. A. Hauck, W. B. Kunkel, D. S. McDonald, and M. D. Williams, *Rev. Sci. Instrum.* 62, 100 (1991).
3. A. T. Young, P. Chen, W. B. Kunkel, K. N. Leung, C. Y. Li, and J. M. Watson, Proc. of the 1991 IEEE Particle Accelerator Conference, San Francisco, CA, May 6 - 9 (1991), p. 1993.
4. D. S. Pickard, K. N. Leung, L. T. Perkins, D. M. Ponce, and T. Young, *Rev. Sci. Instrum.* 67, 428 (1996).
5. K. Saadatmand, G. Arbique, J. Hebert, and R. Valicenti, and K. N. Leung,*Rev. Sci. Instrum.* 66, 3438, (1995) .
6. L. T. Perkins, P. R. Herz, K. N. Leung, and D. S. Pickard, *Rev. Sci. Instrum.*, 67, 10 (1996).
7. C.-F. Chan and K. N. Leung, *Particle Accelerators*, Vol. 43(3), 145 (1994).
8. L. T. Perkins, P. R. Herz, K. N. Leung, and D. S. Pickard, *Rev. Sci. Instrum.*, 65, 1186 (1994).
9. L. T. Perkins (to be published).

10. D. S. Pickard, W. B. Kunkel, K. N. Leung, and A. T. Young, *Rev. Sci. Instrum.*, <u>67</u>, 1666(1996).

Positron Production in Multiphoton Light-by-Light Scattering

presented by

Christian Bula

for the E-144 Collaboration:

C. Bamber[2,*], S.C. Berridge[5], S.J. Boege[2,†], W.M. Bugg[5], C. Bula[1],
D.L. Burke[4], R.C. Field[4], G. Horton-Smith[4], T. Koffas[2], T. Kotseroglou[2,4],
K.T. McDonald[1], A.C. Melissinos[2], D.D. Meyerhofer[2,3], E.J. Prebys[1], W. Ragg[2,§],
D.A. Reis[2], K. Shmakov[5], J.E. Spencer[4], D. Walz[4] and A.W. Weidemann.[5]

[1] *Joseph Henry Laboratories, Princeton University, Princeton, NJ 08544*
[2] *Dept. of Physics and Astronomy,* [3] *Dept. of Mechanical Engineering,*
University of Rochester, Rochester, NY 14627
[4] *Stanford Linear Accelerator Center, Stanford University, Stanford, CA 94309*
[5] *Dept. of Physics and Astronomy, University of Tennessee, Knoxville, TN 37996*

Abstract. A signal of 106 ± 14 positrons above background has been observed in collisions of a low-emittance 46.6-GeV electron beam with terawatt pulses from a Nd:glass laser at 527 nm wavelength in an experiment at the Final Focus Test Beam at SLAC. Peak laser intensities of $\sim 1.3 \times 10^{18}$ W/cm^2 have been achieved corresponding to a value of 0.3 for the parameter $\Upsilon = \mathcal{E}^*/\mathcal{E}_{\rm crit}$ where $\mathcal{E}^* = 2\gamma\mathcal{E}_{\rm lab}$ is the electric field strength of the laser transformed to the rest frame of the electron beam and $\mathcal{E}_{\rm crit} = m^2c^3/e\hbar = 1.3 \times 10^{16}$ V/cm is the QED critical field strength. The positrons are interpreted as arising from a two-step process in which laser photons are backscattered to GeV energies by the electron beam followed by a collision between the high-energy photon and several laser photons to produce an electron-positron pair. These results are the first laboratory evidence for a light-by-light scattering process involving only real photons.

---

* Present address: Hughes Leitz Optical Technologies Ltd., Midland, Ontario, L4R 2H2, Canada
† Present address: Lawrence Livermore National Laboratory, Livermore, CA 94551
§ Present address: Panoramastrasse 8, 78589 Durbheim, Germany

INTRODUCTION

Following the discovery of the positron by Anderson in 1932 [1], Bethe and Heitler [2] provided a theory of the production of electron-positron pairs as arising from the interaction of a real photon with a virtual photon of the electromagnetic field of a nucleus. Shortly thereafter, Breit and Wheeler [3] calculated the cross section for production of an electron-positron pair in the collision of two real photons,

$$\omega_1 + \omega_2 \to e^+ e^-, \qquad (1)$$

to be of order r_e^2, where r_e is the classical electron radius. While pair creation by real photons is believed to occur in astrophysical processes [4] it has not been observed in the laboratory up to the present.

After the invention of the laser in 1960 the prospect of intense laser beams led to reconsideration of the Breit-Wheeler process by Reiss [5] and others [6,7]. Of course, for production of an electron-positron pair the center-of-mass energy of the scattering photons must be at least $2mc^2 \approx 1$ MeV. This can be achieved by scattering a laser beam against a high-energy photon beam created, for example, by backscattering the laser beam off a high-energy electron beam [8]. With laser light of wavelength 527 nm (energy 2.35 eV), a photon of energy 109 GeV would be required for reaction (1) to proceed. However, with an electron beam of energy 46.6 GeV as available at the Stanford Linear Accelerator Center (SLAC) the maximum Compton-backscattered photon energy from a 527-nm laser is only 29.2 GeV.

In strong laser fields the interaction need not be limited to initial states with two photons [5], but rather the number of interacting photons becomes large as the dimensionless, invariant parameter

$$\eta = \frac{e\mathcal{E}_{\rm rms}}{m\omega_0 c} = \frac{e\mathcal{E}_{\rm rms}\lambda_0/2\pi}{mc^2} = \frac{e\sqrt{\langle A_\mu A^\mu \rangle}}{mc^2} \qquad (2)$$

approaches and exceeds unity. In this, the laser beam has laboratory frequency ω_0, wavelength λ_0, root-mean-square electric field $\mathcal{E}_{\rm rms}$, and four-vector potential A_μ; e and m are the charge and mass of the electron, respectively, and c is the speed of light. Thus the multiphoton Breit-Wheeler reaction,

$$\omega + n\omega_0 \to e^+ e^-, \qquad (3)$$

becomes accessible for $n \geq 4$ laser photons of wavelength 527 nm colliding with a photon with $\hbar\omega = 29$ GeV.

For photons of wavelength 527 nm a value of $\eta = 1$ corresponds to laboratory field strength of $\mathcal{E}_{\rm lab} = 6 \times 10^{10}$ V/cm and intensity $I = 10^{19}$ W/cm². Such intensities are now practical in tabletop laser systems based on chirped-pulse amplification [9].

When a laser field of strength $\mathcal{E}_{\text{lab}}$ is viewed in the rest frame of a relativistic, counter-propagating particle with laboratory energy E and Lorentz factor $\gamma = E/mc^2 \gg 1$ the laser field strength appears boosted to $\mathcal{E}^{\star} = 2\gamma\mathcal{E}_{\text{lab}}$. For example, a 46.6-GeV electron has $\gamma = 9 \times 10^4$ so if it collides head on with a 527-nm laser pulse of strength $\eta = 1$ the field in the electron's rest frame is $\mathcal{E}^{\star} = 1.1 \times 10^{16}$ V/cm. This is close to the quantum electrodynamic (QED) critical field strength $\mathcal{E}_{\text{crit}} = m^2c^3/e\hbar = 1.3 \times 10^{16}$ V/cm at which the energy gain of an electron accelerating over a Compton wavelength is its rest energy, and at which a static electric field would spontaneously break down into electron-positron pairs [10-12].

Indeed, the predicted rates [5-7] for reaction (3) become large only when the dimensionless invariant

$$\Upsilon = \frac{\mathcal{E}^{\star}}{\mathcal{E}_{\text{crit}}} = \frac{\sqrt{(F_{\mu\nu}p^{\nu})^2}}{mc^2 \mathcal{E}_{\text{crit}}} \tag{4}$$

approaches unity. Here $F_{\mu\nu}$ is the laboratory electromagnetic field tensor of the laser beam and p^{ν} is the energy-momentum 4-vector of the high-energy electron. For given electron and photon energies E and ω_0 the parameters η and Υ are not independent, and for $E = 46.6$ GeV and $\hbar\omega_0 = 2.35$ eV they are related by $\Upsilon = 0.84\,\eta$.

In reaction (3) where several laser photons interact at once it is useful to consider the interaction as taking place with the field rather than individual quanta. This leads to an interpretation of the pair creation as a barrier-penetration process. A virtual electron-positron pair in the vacuum can materialize if the charges separate by distance d sufficient to extract energy $2mc^2$ from the field, i.e. if $e\mathcal{E}d = 2mc^2$. The probability of penetration of this 'barrier' of thickness d is proportional to $\exp(-2d/\lambda_C) = \exp(-4m^2c^3/e\hbar\mathcal{E}) = \exp(-4/\Upsilon)$, where λ_C is the Compton wavelength of the electron. A more complete calculation of this process [10-12] indicates that the rate for pair production $(R_{e^+e^-})$ is

$$R_{e^+e^-} \propto \exp(-\pi/\Upsilon). \tag{5}$$

In addition to pursuing the basic physics program outlined above, our experiment provides a demonstration of the technology for e-γ and γ-γ collider options [13], leading to measurements of the γWW coupling via the reaction $e\gamma \to W\nu$ [14,15], etc. Also, copious production of positrons in e-γ collisions could provide a low-emittance positron source due to the absence of final-state Coulomb scattering [16].

EXPERIMENTAL SETUP

We have performed an experimental study of strong-field QED in the collision of a 46.6-GeV electron beam with terawatt pulses from a frequency doubled Nd:glass laser. A schematic diagram of the experiment is shown in Fig. 1. The apparatus was designed to detect electrons that undergo nonlinear Compton scattering,

$$e + n\omega_0 \to e' + \omega, \qquad (6)$$

as well as positrons from the two-step process of reaction (6) followed by reaction (3). Measurements of reaction (6) have been reported elsewhere [17,18].

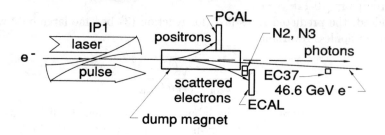

FIGURE 1: Schematic layout of the experiment.

The experiment was carried out in the Final Focus Test Beam (FFTB) at SLAC [19]. The laser beam was focused onto the electron beam by an off-axis parabolic mirror of 30-cm focal length with a 17° crossing angle at the interaction point, IP1, 10 m downstream of the Final Focus.

The laser was a 1.5-ps (fwhm), chirped-pulse-amplified Nd:glass terawatt system with a relatively high repetition rate of 0.5 Hz achieved by a final laser amplifier with slab geometry [18,20,21]. The laser-oscillator mode locker was synchronized to the 476-MHz drive of the SLAC linac klystrons with an observed jitter between the laser and linac pulses of 2 ps (rms) [22]. The spatial and temporal overlap of the electron and laser beams was optimized by observing the Compton scattering rate in the EC37, N2, N3 and ECAL detectors during horizontal, vertical, and time scans of one beam across the other [21].

The intensity of the laser at the focus was determined from measurements of the laser energy, focal-spot area, and pulse width. The uncertainty in the pulse width was ±35% in that measurements could be made only occasionally with a single-shot autocorrelator. Fluctuations on the energy probe calibration led to a ±20% uncertainty in the energy measurement. The focal spot area at IP1 was measured by reimaging the focus of the laser on a CCD. Because of laser light scattering, filtering, and a non-Gaussian shape of the focal spot the uncertainty in the area was ±30%. The overall uncertainty in the laser intensity as determined by these diagnostic devices was therefore ±50%.

The peak focused laser intensity was obtained for green pulses of energy $U = 650$ mJ, focal area $A \equiv 2\pi\sigma_x\sigma_y = 30$ μm^2, and pulse width $\Delta t = 1.6$ ps (fwhm), for which $I = U/A\Delta t \approx 1.3 \times 10^{18}$ W/cm^2 at $\lambda_0 = 527$ nm, corresponding to values of $\eta = 0.36$ and $\Upsilon = 0.3$.

The electron beam was operated at 10-30 Hz with an energy of 46.6 GeV and emittances $\epsilon_x = 3 \times 10^{-10}$ m-rad and $\epsilon_y = 3 \times 10^{-11}$ m-rad. The beam was tuned to a focus with typically $\sigma_x = 25$ μm and $\sigma_y = 40$ μm at the laser-electron interaction point. The electron bunch length was expanded to 3 ps (rms) to minimize the effect of the time jitter between the laser and electron pulses. Typical bunches contained 7×10^9 electrons. However, since the electron beam was significantly larger than the laser focal area only a small fraction of the electrons crossed through the peak field region.

A string of permanent magnets after the collision point deflected the electron beam downwards by 20 mrad. Electrons and positrons of momenta less than 20 GeV were deflected by the magnets into two Si-W calorimeters (ECAL and PCAL) as shown in Fig. 1. The calorimeters were made of alternating layers of silicon (300 μm) and tungsten (one radiation length) and measured electromagnetic shower energies with resolution $\sigma_E/E \approx 19\%/\sqrt{E[\text{GeV}]}$ (plus a constant electronic noise of 250 MeV). Each layer of silicon was divided into horizontal rows and 4 vertical columns of 1.6×1.6 cm^2 active area cells, which allowed the determination of isolated shower positions with resolution of 2 mm.

The Si-W calorimeters were calibrated in parasitic running of the FFTB to the SLC program in which linac-halo electrons of energies between 5 and 25 GeV were transmitted by the FFTB when the latter was tuned to a lower energy. The number of such electrons varied between 1 and 100 per pulse, which provided an excellent calibration of the ECAL and PCAL over a wide dynamic range. The calibration runs also confirmed the magnetic-field maps of the FFTB dump magnets that are used in our spectrometer.

Electrons scattered via reaction (6) for $n = 1$, 2 and 3 laser photons were measured in gas Čerenkov counters labeled EC37, N2 and N3 in Fig. 1. These counters were used to monitor the quality of the e-laser beam overlap and to extract the field intensity at the laser focus on each shot. We used detectors based on Čerenkov radiation because of their insensitivity to major sources of low-energy background, such as beam scraping and (in the case of N2 and N3) recoil electrons produced by Compton scattered electrons hitting beamline components. EC37 was calibrated by inserting a thin foil in the electron beam at IP1. The momentum acceptance and efficiency of the counters N2 and N3 were measured with the parasitic electron beam by comparison with the previously calibrated ECAL.

RESULTS

We used the PCAL calorimeter to search for positrons produced at IP1. Because of the high rate of electrons in the ECAL calorimeter from Compton scattering it was not possible to identify the electron partners of the positrons.

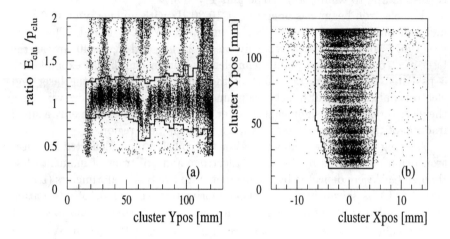

FIGURE 2: Cluster densities from Bethe-Heitler positrons produced by a wire at IP1. The solid line shows the signal region for positron candidates. (a) Ratio of cluster energy to momentum $vs.$ vertical impact position. The low ratios at the center of PCAL are caused by a 1.5-mm-wide inactive gap. Similarly, at the top and bottom of PCAL a part of the shower energy is lost due to leakage out of PCAL. Two simultaneous showers separated by less than a cell caused the clusters with $E_{\text{clu}}/P_{\text{clu}} \sim 2$. (b) Cluster position in PCAL.

The response of PCAL to positrons originating at IP1 was studied by inserting a wire at IP1 to produce Bethe-Heitler e^+e^- pairs. These data were used to develop an algorithm to group contiguous PCAL cells containing energy deposits into 'clusters' representing positron candidates. The clusters were characterized by their position in the horizontal (X_{pos}) and vertical (Y_{pos}) direction and their total energy deposit E_{clu}. Using the field maps of the magnets downstream of IP1, the vertical impact position was translated into the corresponding momentum P_{clu} which could be compared to the cluster energy. Fig. 2 shows the density of clusters produced by the wire in the two planes $E_{\text{clu}}/P_{\text{clu}}$ $vs.$ Y_{pos} and Y_{pos} $vs.$ X_{pos}. Only clusters within the signal regions bounded by solid lines in Fig. 2 were counted as positron candidates.

We collected data at various laser intensities. The data from collisions with poor e-laser beam overlap were discarded. Also, events with anomalous values

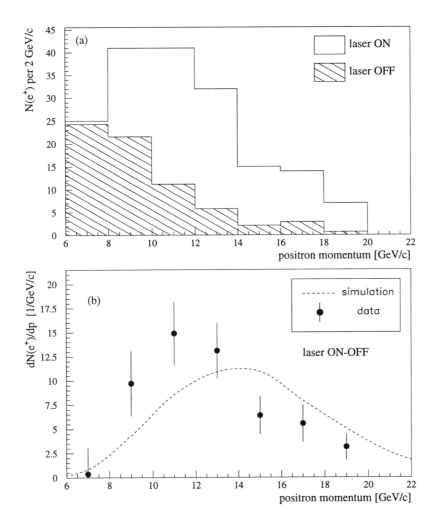

FIGURE 3: (a) Number of positron candidates *vs.* momentum for laser-on pulses and for laser-off pulses (hatched distribution, scaled to the number of laser-on pulses). (b) Spectrum of signal positrons obtained by subtracting the laser-off from the laser-on distribution. The dashed line shows the expected momentum spectrum from the model calculation. PCAL cluster positions have been converted to positron momentum via knowledge of the field in the magnetic spectrometer.

for any of the measured electron or laser beam parameters were removed from the data sample. The number of positron candidates observed in the remaining 21,962 laser shots is 175±13 and is shown as the upper distribution in Fig. 3(a) as a function of cluster momentum.

Positrons were also produced in showers of lost electrons upstream of the e-laser interaction point. The rate of these background positrons was studied in 121,216 electron-beam pulses when the laser was off, yielding a total of 379 ± 19 positron candidates. Fig. 3(a) shows the momentum spectrum of these candidates as the hatched distribution, which has been scaled by 0.181, this being the ratio of the number of laser-on to laser-off pulses. After subtracting the laser-off distribution from the laser-on distribution we obtain the signal spectrum shown in Fig. 3(b) whose integral is 106 ± 14 positrons. The statistical significance of this result, by itself, is in excess of seven standard deviations. Even more significantly the momentum distribution of the observed positrons and the dependence of the rate on the laser intensity confirm that the positrons originate from light-by-light scattering, as discussed below.

We have modeled the pair production as the two-step process corresponding to reaction (6) followed by reaction (3). We followed the formalism of Ref. [6] for linearly polarized light as used in the experiment. By numerical integration over space and time in the e-laser interaction region we account for both the production of the high-energy photon (through a single or multiphoton interaction) and its subsequent multiphoton interaction within the same laser focus to produce the pair. Further Compton scatters of the positron (or electron) are also taken into account. The positron spectrum predicted by this calculation is shown as the dashed line in Fig. 3(b) and is in reasonable agreement with the data.

As mentioned before, several laser photons are needed to produce an e^+e^- pair under the present experimental conditions. The numerical simulation of the two-step Breit-Wheeler process, (6) followed by (3), indicates that the average number n of photons absorbed from the laser field in the second step is between 4 and 5 for a peak field intensity $\Upsilon \leq 0.35$. Fig. 4 shows the probability distribution of n for $\Upsilon = 0.3$ at the laser focus.

For an additional determination of the laser intensity we made use of N_1, N_2 and N_3, the numbers of electrons intercepted by the gas Čerenkov counters EC37, N2 and N3, of first-, second- and third-order Compton scattering, respectively. In principle, the field intensity could be extracted from each of these monitors. However, the result is more stable against various experimental uncertainties such as e-laser timing jitter if it is extracted only from ratios of the monitor rates. For $\eta^2 \ll 1$, the field intensity is approximately given by

$$\eta^2 = k_1 \cdot \frac{N_2}{N_1}, \quad \text{and} \quad \eta^2 = k_2 \cdot \frac{N_3}{N_2}. \tag{7}$$

The parameters k_1 and k_2 depend on the acceptance and efficiency of the

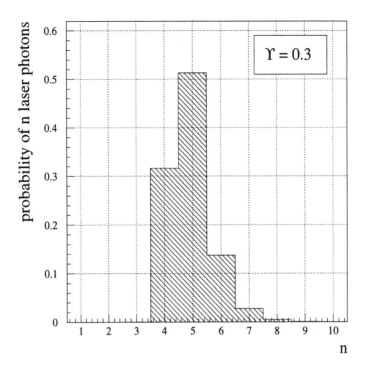

FIGURE 4: Calculated probability distribution of the number n of photons absorbed from the laser field in the second step of the two-step Breit-Wheeler pair creation process. A field intensity of $\Upsilon = 0.3$ at the laser focus was used for the simulation.

counters as well as the spectrum of scattered electrons and were calculated over the relevant range of η^2 values by the numerical simulation. We fit the observed N_i to ideal values subject to the constraint $N_2^2 = (k_2/k_1)N_1 N_3$ obtained from Eq. (7). Then the fitted N_i were used to determine η and Υ for each laser shot with an average precision of 13%. Uncertainties in the acceptance and efficiency of the counters caused a systematic error of $\sim 20\%$ to the absolute value of η and Υ. The intensity at the laser focus deduced by this method is in good agreement with the average value calculated from the measured laser parameters.

Fig. 5 shows the yield of positrons/laser shot (R_{e^+}) as a function of Υ. The solid line is a power law fit to the data and gives

$$R_{e^+} \propto \Upsilon^{10.0 \pm 0.4 \,(\text{stat.}) \pm 0.4 \,(\text{syst.})}, \tag{8}$$

where the statistical error is from the fit and the systematic error was estimated

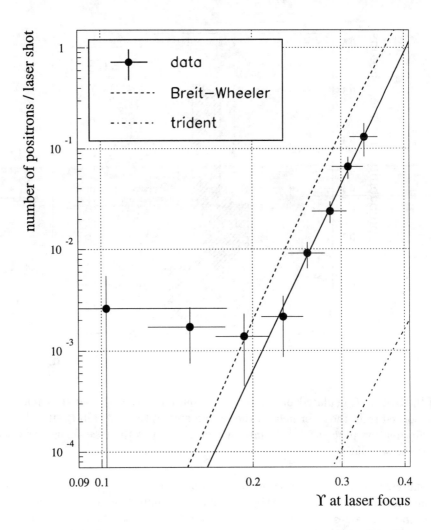

FIGURE 5: Dependence of the positron rate on the laser intensity. The solid line shows a power law fit to the data. The dashed line is the prediction based on the numerical integration of the two-step Breit-Wheeler process, (6) followed by (3). The shift between the data and this simulation is well within the combined effect of the systematic uncertainty of 45% in the e-laser overlap efficiency and the 20% uncertainty in the absolute value of Υ. The dash-dot line represents the calculation for the one-step trident process (10) with an intermediate virtual photon.

by choosing different bin sizes in Υ. Thus, the observed positron production rate is highly nonlinear, varying as the 10th power of the electric field strength. This is in good agreement with expectations as on average $n = 5.5$ photons are needed to produce a pair (1 in reaction (6) and 4.5 in (3)) and the rate of multiphoton reactions involving n laser photons is approximately proportional to Υ^{2n}. Several points at low values of Υ seen in Fig. (5), while statistically consistent with the fit in Eq. (8), indicate a possible residual background of $\sim 2 \times 10^{-3}$ positrons/laser shot in the data sample.

The dashed curve in Fig. 5 shows the prediction based on the numerical integration of the two-step Breit-Wheeler process, (6) followed by (3), and confirms the observed rate dependence on Υ. The simulated rate has been reduced by a factor of 0.35 to account for the average efficiency in e-laser overlap of $35\% \pm 15\%$ as deduced from the Compton monitors EC37, N2 and N3. The apparent shift between the data and this simulation is well within the combined effect of the systematic uncertainty in the e-laser overlap efficiency and the 20% uncertainty in the absolute value of Υ.

To confirm the form of Eq. (5) we plot the yield of positrons/laser shot (R_{e+}) as a function of $1/\Upsilon$ in Fig. 6. The solid line is an exponential fit to the data and gives

$$R_{e+} \propto \exp[(-2.8 \pm 0.2 \,(\text{stat.}) \pm 0.2 \,(\text{syst.}))/\Upsilon], \tag{9}$$

with a χ^2 per degree of freedom of 1.13. This result is in close agreement with the prediction of Eq. (5).

Although we have demonstrated a signal of positron production associated with scattering of laser light we cannot immediately distinguish positrons from reaction (3) from those originating in the trident process

$$e + n\omega_0 \to e'e^+e^-, \tag{10}$$

which is the Bethe-Heitler process for an electron target. A complete theory of reaction (10) does not exist at present so we have performed calculations based on a two-step model in which the beam electron emits a virtual-photon according to the Weizsäcker-Williams approximation and the virtual photon combines with laser photons to yield electron-positron pairs according to the theory of the multiphoton Breit-Wheeler process (3). This is distinct from the real-photon calculation previously discussed. The results of this simulation indicate that for the interaction geometry of the present experiment and the values of Υ achieved, the trident process is suppressed by more than three orders of magnitude. The expected trident rate, also corrected for e-laser overlap efficiency, is shown in Fig. 5 as the dash-dot line.

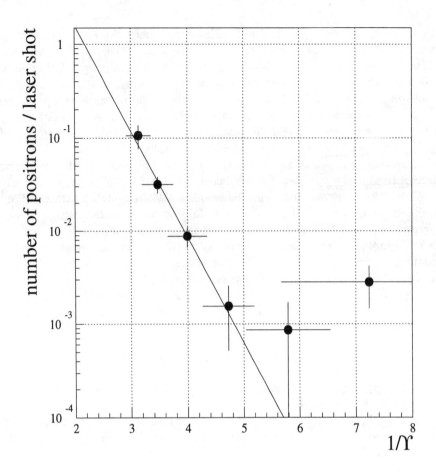

FIGURE 6: Positron yield per laser shot *vs.* $1/\Upsilon$. The solid line shows an exponential fit to the data and confirms the form predicted by Eq. (5).

CONCLUSION

These results, as well as those presented in Ref. [17], confirm the validity of the formalism of strong-field QED and show that the observed rates for the multiphoton reactions (3) and (6) are in agreement with the predicted values. Furthermore these results are the first demonstration of breakdown of the vacuum by an intense electromagnetic wave, and they are the first observation of photon-photon scattering with real photons.

ACKNOWLEDGMENTS

We thank the SLAC staff for its extensive support of this experiment. The laser system could not have been completed without support from members of the Laboratory for Laser Energetics at U. Rochester. T. Blalock was instrumental in the construction of the laser system and its installation at SLAC. We also thank U. Haug, A. Kuzmich and D. Strozzi for participation in recent data collection, and A. Odian and P. Chen for many useful conversations. KTM wishes to thank J.A. Wheeler for continued inspiration. This work was supported in part by DoE grants DE-FG02-91ER40671, DE-FG02-91ER40685, DE-FG05-91ER40627 and contract DE-AC03-76SF00515.

REFERENCES

1. C.D. Anderson, Science **76**, 238 (1932); Phys. Rev. **43**, 491 (1933).
2. H.A. Bethe and W. Heitler, Proc. Roy. Soc. **A146**, 83 (1934).
3. G. Breit and J.A. Wheeler, Phys. Rev. **46**, 1087 (1934).
4. O.C. De Jager et al., Nature **369**, 294 (1994).
5. H.R. Reiss, J. Math. Phys. **3**, 59 (1962).
6. A.I. Nikishov and V.I. Ritus, Sov. Phys. JETP **19**, 529, 1191 (1964); **20**, 757 (1965).
7. N.B. Narozhny et al., Sov. Phys. JETP **20**, 622 (1965).
8. R.H. Milburn, Phys. Rev. Lett. **10**, 75 (1963).
9. D. Strickland and G. Mourou, Opt. Comm. **55**, 447 (1985).
10. F. Sauter, Z. Phys. **69**, 742 (1931).
11. W. Heisenberg and H. Euler, Z. Phys. **98**, 718 (1936).
12. W. Greiner and J. Reinhardt, *Quantum Electrodynamics*, Berlin, Springer, 1992, p. 285.
13. Nucl. Instr. and Meth. **355**, (1995).
14. K.O. Mikaelian, Phys. Rev. D **17**, 750 (1978); **30**, 1115 (1984).
15. I.F. Ginzburg et al., Nucl. Phys. **B228**, 285 (1983).
16. P. Chen and R.B. Palmer, SLAC-PUB-5966 (Nov. 1992).
17. C. Bula et al., Phys. Rev. Lett. **76**, 3116 (1996).
18. T. Kotseroglou, Ph.D. thesis, U. Rochester, UR-1459 (Jan. 1996).
19. V. Balakin et al., Phys. Rev. Lett. **74**, 2479 (1995).
20. C. Bamber et al., U. Rochester preprint UR-1428 (June 1995).
21. S.J. Boege, Ph.D. thesis, U. Rochester, UR-1458 (Jan. 1996).
22. T. Kotseroglou et al., Nucl. Instr. and Meth. A **383**, 309 (1996).

Inverse Free Electron Laser Acceleration with a Square Wave Wiggler

Z. Parsa[†1] and M.P. Pato[‡2]

[†]*Brookhaven National Laboratory*
901A Physics Dept., Upton, NY 111973-5000, USA
[‡]*Instituto de Fisica, Universidade de Sao Paulo*
C.P.20516,01498 Sao Paulo,S.P., Brazil

Abstract. We present an Inverse Free Electron Laser with a Square Wave Wiggler (IFELSW) as a new acceleration scheme and show Analytically and numerically about factor of 2 gain in the energy when compared to the standard IFEL with the Sinusoidal [1] field Wiggler.

I INTRODUCTION

The Nonlinear Amplification of Inverse-Bremsstrahlung Electron Acceleration (NAIBEA) is a scheme of acelerating charged particles that uses a laser coupled to a static applied field structure in which a constant magnetic, or electric, field alternates sign at some appropriately determined positions in such a way that the particle is always accelerated. This may be understood as a kind of the Inverse Free Electron Laser Acceleration (IFEL). In both of these acceleration schemes [1,2], relativistic particles move under the combined action of an electromagnetic travelling wave, the laser field, and a magnetic, or electric, applied static field. Both, the laser radiation and the beam of particles, propagate along the field structure. This static field, usually provided by magnets, acts like a wiggler by producing a small undulation in the particle trajectory. The transverse velocity of this undulation couples in such a way to the electric field of the electromagnetic wave that the energy is transferred from the laser to the beam.

In the standard IFEL this wiggling motion is created by an undulator whose magnetic field varies sinusoidally with the longitudinal distance [1], while in

---

[1)] Supported by US Department of Energy Contract No. DE-AC02-76CH00016 and National Science Foundation NSF-PHY-94-07194
[2)] Supported in part by the CNPq-Brazil

our new IFEL scheme the wiggling motion is created by a square wave wiggler (IFELSW). Similarly, in our new NAIBEA scheme the applied field is taken as a constant field that changes sign abruptly and therefore it behaves more like a square wave. We show and explore the new NAIBEA scheme as an IFEL with a square wave wiggler(IFELSW). To do this we introduce some modifications in the NAIBEA [2] field structure and in the way the electrons enter it. In this paper we show both analytically and numerically how (IFELSW) IFEL with a square wave wiggler (and the new NAIBEA) accelerates particles.

In previous calculations, the positions where the field switch sign were determined and adjusted numerically and the electrons were supposed to be injected in the laser field with a small transversal velocity component [2]. This later circumstance introduces an initial asymmetry in the field structure that does not favor the capacity of accelerating an electron beam. We show that by constructing a structure (such as a wiggler) in which the sign is switched exactly when the electron reaches the points where the phase are odd numbers of $\frac{\pi}{2}$, it is accelerated without having any initial transversal velocity. Which certainly makes the new scheme more efficient.

Thus We define now the applied field of the wiggler to be exactly a square wave when it is considered as a function of the phase of the electric laser field at the position where the particle is. This square wave is assumed to have period of 2π and be shifted by $\frac{\pi}{2}$ with respect to the laser field. The electrons are to be injected along with the laser radiation, (with no transversal velocity). This will remove the initial asymmetry in the NAIBEA [2] field structure, enhances the capacity of accelerating an electron beam and will permit to study the effectiveness of the square wave structure as compared to the IFEL sinusoidal wiggler. One important property of IFEL is the fact that it produces as a final result a bunched beam of accelerated electrons. This effect is caused by the stability of the resonant condition that creates a potential well in which particles around this optimum situation get trapped and are dragged to high energies [1]. Numerical simulations suggest that our (IFELSW) square well IFEL scheme has this same property.

II THE RESONANT TRAJECTORY

We consider the movement of an electron under the action of a linearly plane polarized laser electromagnetic (EM) wave and an applied transverse magnetic field of the wiggler ($\vec{B}_{app}$). Assuming the wave is propagating in the z-direction, see Fig.1, the relativistic equation of motion

$$\frac{d\vec{P}}{dt} = -e\left[\vec{E}_{laser} + \vec{\beta} \times (\vec{B}_{laser} + \vec{B}_{app})\right] \quad (1)$$

where $\vec{\beta} = \frac{\vec{v}}{c}$, can be expressed in terms of its components as

$$\frac{dP_z}{dt} = -e\beta_x B_y - e\beta_x B_{app} \tag{2}$$

$$\frac{dP_x}{dt} = -eE_x + e\beta_z B_y + e\beta_z B_{app}. \tag{3}$$

With $E_x = B_y = -E_{x_0}\sin\phi$, $\phi = k(ct-z)$ and $B_{app} = B_0 f(\phi)$, where ϕ corresponds to the phase of the laser field at the particle position, the above equations become

$$\frac{dP_z}{dt} = e\beta_x E_{x_0}\sin\phi - e\beta_x B_0 f(\phi) \tag{4}$$

$$\frac{dP_x}{dt} = e(1-\beta_z)E_{x_0}\sin\phi + e\beta_z B_0 f(\phi) \tag{5}$$

where $f(\phi)$ is a unit square wave of period 2π that switch signs at odd numbers of $\frac{\pi}{2}$ and $f(0) = -1$. By multiplying the above equations by v_z and v_x, respectively, we obtain the energy equation

$$mc^2\frac{d\gamma}{dt} = ev_x E_{x_0}\sin\phi. \tag{6}$$

Eqs. (4) and (6) can be combined to give

$$c\frac{dP_z}{dt} - mc^2\frac{d\gamma}{dt} = -ev_x B_0 f(\phi). \tag{7}$$

First integrals of the motion are obtained by integrating Eqs. (4) and (5)

$$P_z c = mc^2\gamma - e\int_0^t dt v_x B_0 f(\phi) + K_1 \tag{8}$$

$$P_x = -eE_{x_0}\cos\phi - e\int_0^t dt v_z B_0 f(\phi) + K_2 \tag{9}$$

where we used the fact that $\frac{d\phi}{dt} = kc(1-\beta_z)$. These two integrals can be performed by parts using the fact that since $f(\phi)$ is a square wave function its derivative is a sum of delta functions, namely,

$$f'(\phi) = 2\sum_n (-1)^n \delta\left[\phi - (2n+1)\frac{\pi}{2}\right]. \tag{10}$$

We obtain

$$P_z c = mc^2\gamma - eB_0 x(\phi) f(\phi) + K_i \tag{11}$$

$$P_x = -eE_{x_0}\cos\phi - eB_0 z(\phi)f(\phi) - 2eB_0 \sum_{n=1}^{i}(-1)^n z_n + K_2 \qquad (12)$$

where

$$K_i = -K_1 - eB_0 x_0 + 2B_0 \sum_{n}^{i}(-1)^n x_n \qquad (13)$$

and i is the running index corresponding to the segment of the wiggler where the particle is located.

In reducing the above system of equations to a one dimensional problem, for convenience we first change the independent variable from the time (t) to (ϕ), and use a new variable Q such that $\frac{dQ}{d\phi} = P_x$, i.e.,

$$Q' = P_x = m\gamma\frac{d\phi}{dt}x'(\phi) = \frac{k}{c}(mc^2\gamma - P_z)x'(\phi) \qquad (14)$$

Where the primes denote derivatives with respect to ϕ and we have used the relation $\frac{d\phi}{dt} = k(c - v_z) = \frac{k}{mc\gamma}(mc^2\gamma - P_z)$. Substituting P_x in Eq. 14 results in

$$Q' = -\frac{k}{c}B_0 x(\phi)x'(\phi)f(\phi) - \frac{k}{c}K_i x'(\phi). \qquad (15)$$

and after integrating we obtain

$$Q = -\frac{k}{2c}eB_0 x^2(\phi)f(\phi) - \frac{k}{c}K_i x(\phi) + C_i \qquad (16)$$

where the constant C_i is introduced in order to assure the continuity of coordinate Q when the particle traverses the switching sign points of the field. They are given by

$$C_i = \frac{eB_0}{kc}\sum_{n}^{i}(-1)^n(kx_n)^2 + \frac{K_i k x_0}{c} \qquad (17)$$

Since $Q'' = P'_x$, Eq. (9) and relativistic relations can be used to eliminate the dependence of Q in other variables except ϕ (the phase of the laser field). This leads to the following equation of motion for Q as a function of ϕ only:

$$Q'' = -\frac{eB_0 c^2}{2}f(\phi)\frac{m^2c^2 + Q'^2}{kK_i^2 - 2ceB_0 f(\phi)(Q - C_i)} + eB_{app}f(\phi) + eE_{x_0}\sin\phi \qquad (18)$$

This is a nonlinear second order differential equation. It can be used to predict the existence of a periodic solution as illustrated In Fig.3, We have the evolution of the variable Q and of the transversal displacement x, obtained inverting Eq. (16), as a function of ϕ. This is a resonant trajectory. It is locally periodic, in the sense that its amplitude and period vary slowly as the particle accelerates.

We observe in Fig. 3 that the electron enters and leaves each segment (in which the applied field has a given sign), with $x = 0$ which means that the constants K_i and C_i (given by Eqs. (13) and (17)) remain the same in all segments. We further note, that multiplying the equation (18) by $f(\phi)$ the product $f(\phi)Q(\phi)$ satisfies an equation which remains invariant, since $f^2(\phi) = 1$ and $f(\phi)\sin\phi$ does not change from one segment to the other. As a consequence an oscilatory periodic motion for Q follows. The relatively small perturbation to the periodicity being caused by the lack of symmetry of the laser field term (the last term in Eq.18) inside each segment of the square wave.

A solution to this nonlinear second order differential equation can be constructed and used to obtain an expression for the rate of energy increase. However here, we follow a simpler approximate approach by obtaining from Eq. (12), an expression for the transversal velocity v_x and puting it into the Eq. (6) (neglecting the oscilatory parts, as we are not interested in). We then observe that in a given segment of the applied field, the variation of the longitudinal distance z is exactly half of the wiggler period Λ_w. Since z is a monotonic function of ϕ, (in the average) we can further assume that z increases linearly with ϕ. Then the two terms left in the expression of v_x define a triangular wave. So we can write,

$$v_x \sim \frac{eB_0\Lambda_w}{4mc^2\gamma}g(\phi) \qquad (19)$$

where $g(\phi)$ is a unit triangular wave with the Fourier representation

$$g(\phi) = -\sum_{n=1}^{\infty}\frac{8}{n^2\pi^2}\sin\frac{n\pi}{2}\sin n\phi \qquad (20)$$

Substituting this into Eq.19 and equation 19 into Eq.6 (again neglecting the high frequency terms) we find,

$$\frac{d\gamma}{dt} = \frac{eE_{x_0}}{mc^2}\frac{eB_0\Lambda_w}{8\gamma mc^2} \qquad (21)$$

This expression is to be compared (as in e.g. [10]) with the equivalent one for a sinusoidal wiggler and a plane polarized wave (Eq. (27) of Ref. [1]). We find that the square wiggler gives a rate of increase that is about two times larger.

III CONCLUSION

We found in the last section, that it is possible to construct an Inverse Free Electron Laser (IFEL) with a square wave wiggler (IFELSW) such that the particle in a resonant trajectory is always accelerated. The particles are supposed to be injected along with the laser radiation, i.e., with no transversal velocity. By giving to the square wave a phase shift of $\frac{\pi}{2}$ the wiggling motion acquired by the electrons will be in phase with the laser field and they will be accelerated.

It was possible to obtain an explicit expression for the average rate of energy increase and as a result we have found a gain in energy of about two times when compared with the standard IFEL with a sinusoidal wiggler. We note that, although Maxwell's equations do not allow an abrupt change of sign of the field, the results we have obtained here suggest that the acceleration scheme using laser field turns out to be more efficient as we approach the square wave field patterns.

REFERENCES

1. E.D. Courant, C. Pellegrini and W. Zakowicz, Phys.Rev. **A32**, 2813 (1985).
2. M.S. Hussein and M.P. Pato, Phys. Rev. Lett. **68**, 1136 (1992).
3. M.P. Pato, M.S. Hussein, and A.K. Kerman, Nucl. Inst. & Meth. in Phys. Res.**A328**, 342 (1993).
4. M.S. Hussein, M.P. Pato and A.K. Kerman, Phys. Rev. **A46**, 1136 3562 (1992).
5. M.S. Hussein, M.P. Pato, Mod. Phys. Lett. **B6**, 747 (1992).
6. M.S. Hussein and M.P. Pato, Int. Jour. of Mod. Phys. **A8**, 3235 (1992).
7. R.M.O. Galvão, M.S. Hussein, M.P. Pato, and A. Serbeto, Phys.Rev. **E49**, R4807 (1994).
8. M. P. Pato, Private communication.
9. Z. Parsa, 'An Inverse Free Electron Laser with a Square Wave Wiggler', Propsal to ATF Steering Committee, Report - October 30, 1996.
10. Z. Parsa, 'Square Wave Wiggler IFEL' Supporting Information for ATF Steering Committee Review Meeting, Report - November 20, 1996; and references therein (1996);
11. Z. Parsa, 'Inverse Free Electron Laser Acceleration Scheme with a Square Wave Wiggler', Presentation UCSB/ITP-96, Santa Barbara, CA (1996).
12. Z. Parsa 'New IFEL' Experimental Proposal for BNL-ATF (1997).
13. Z. Parsa, 'Improved NAIBEA and IFEL', APS-APR97 No. 5933, Presentation Washington D.C. (1997).
14. Z. Parsa, T. Marshall, 'Enhanced IFEL Experiment using a Novel Wiggler' - PAC97 Presentation, Vancouver, Canada (May 1997).

Figure Captions:
Fig.1- Show the directions of the laser field and Injection of Electrons.
Fig.2- A sketch of the Wiggler Magnetic field.
Fig.3- Show the trajectories as the electron moves through the Wiggler.
Fig.4- The evolution of energy as the electron moves through the Wiggler.
Fig.5- Energy distribution in terms of γ.

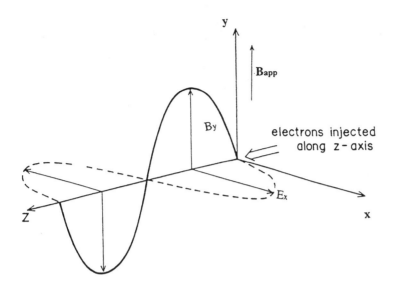

Fig.1-: **Plot of the directions of the laser field and injection of electrons.**

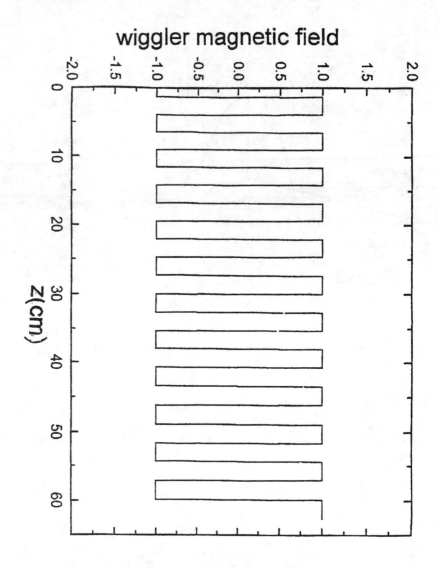

Fig.2- : A sketch of the Wiggler Magnetic field for the proposed IFELSW and the positions where the sign of the magnetic field is to be switched

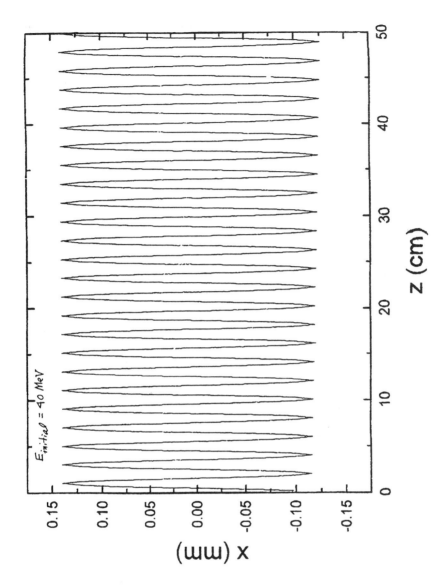

Fig.3- : Plot of the trajectories as the electron moves through the Wiggler (case a Ref. [9])

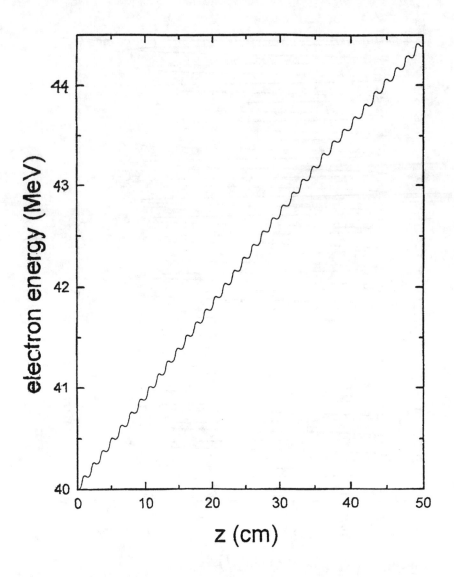

Fig.4- : **Plot showing the predicted IFELSW evolution of the energy as the electron moves through the Wiggler (case a, where the initial prameters are the same as recent BNL- ATF IFEL experiment (case a Ref. [9])**

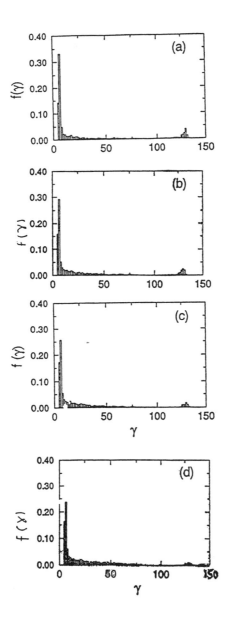

Fig.5- : Plot showing Energy distribution in terms of γ as a normalized histogram for different Temperatures of 10-150 KeV for example of the initial density of 10**8 cm**-3 . Other simulations can be done.

a) T=10 KeV,   b) T=50 KeV,   c) T=100 KeV,   d) T=150 KeV.

New Modes of Particle Acceleration Techniques and Sources

August 19–23, 1996

Coordinator: Z. Parsa

Schedule

Monday, August 19, 1996:
Convener: Z. Parsa

Time	Speaker	Title
8:00 am	Registration	ITP Lobby
8:40	J. Hartle, ITP Director	Welcoming Remarks
	Z. Parsa, BNL	Introduction to Program

Defining Perspective Presentation:

9:10	R. Siemann, SLAC	Status and Future Direction of Advanced Accelerator Research
10:10	Refreshment Break	ITP Front Patio

Convener: A. Sessler
Power Sources - Status, Advances and Limitations:

10:30	R. Phillips, SLAC	RF Sources
11:15	G. Mourou, U. Michigan	Laser as a Power Source
12:00 pm	Lunch Break	ITP Front Patio

Convener: A. Skrinsky

1:30	M. Gunderson, USC	Pulsed Power Sources – Physics Issues Underlying Power Conditioning for Accelerators

Advanced Accelerator Schemes:

2:15	W.B. Mori, UCLA	Laser Acceleration
3:00	Refreshment Break	ITP Front Patio

3:20	S. Yu, LBL	Two Beam Accelerator
4:05	C. Pellegrini, UCLA	Advances in Inverse Free Electron lasers and an Update on Free Electron Lasers
4:50	W. Kimura, STI Optronics	New Advances in Inverse Cerenkov Accelerator
5:30	Wine & Cheese	ITP Front Patio
6:00	Buffet Dinner	ITP Front Patio

Tuesday, August 20, 1996
Convener: C. Pellegrini

Time	Speaker	Title
New Modes of Acceleration & Applications:		
9:00 am	T. Katsouleas, USC	Plasma Beat Wave Accelerator (PBWA) and Overview of Plasma Acceleration
9:45	A. Skrinsky, BINP	Plasma Wakefield Accelerator (PWFA) and System Designs
10:30	Refreshment Break	ITP Front Patio
10:50	C. Joshi, UCLA	Progress in Laser Plasma Experiments
11:35	R. Pantel, Stanford Univ.	Open Waveguide Structure for Laser Acceleration
12:15 pm	Lunch	ITP Front Patio

Convener: R. Siemann

Emittance Issues of Acceleration Methods:		
1:45	E. Esarey, Naval Res. Lab.	Laser Plasma Physics Issues
2:30	N. Andreev. RAS	Self-Modulation of Intense Laser Pulses in Homogeneous Plasma and Plasma Channels
3:15	Refreshment Break	ITP Front Patio
3:35	B. Breizman, Univ. Texas	Beam Dynamic Issues in Plasma Wakefield
4:20	T. Katsouleas, USC	Beam Dynamic Issues in Laser Plasma Accelerators
5:15	**end of session**	

Wednesday, August 21, 1996
Convener: W. B. Mori

Time	Speaker	Title
9:00 am	C. Barty, UCSD	Production of Ultra Short Laser Pulses
9:45	D. Umstadter, U. Michigan	Short Bunch Injection, Synchronization and Acceleration in Laser Wakefields
10:30	Refreshment Break	ITP Front Patio
10:50	A. Skrinsky, BINP	Beam Cooling Techniques
11:35	A. Sessler, LBL	Laser Cooling
12:15 pm	Lunch	ITP Front Patio

Convener: E. Esarey

Time	Speaker	Title
1:45	P. Chen, SLAC	Crystal Accelerator (Plasma Wake in Conduction Electron and Channeling
2:20	A. Skrinsky, Z. Parsa N. Andreev, B. Breizman C. Joshi, W.B. Mori G. Stupakov, T. Katsouleas V. Gorev, D. Umstadter and others	**Reports* on New Advances in Lasers,** Plasma based Acceleration Techniques **and Panel Discussions**
3:20	Refreshment Break	ITP Front Patio
3:40	G. Stupakov	Beam Emittance in PWFA
4:10	T. Marshall, Columbia	Microwave Inverse Cerenkov and FEL Accelerators: "Tabletop" Systems for Applications at 10-20 MeV
4:50	V. Telnov, BINP	Requirements of Beam Emittances in Photon-Photon Colliders and Methods of Realization
5:30	**end of session**	

Thursday, August 22, 1996
Convener: D. Rule

Time	Speaker	Title
9:00 am	A. Shchagin, KIPT	Parametric X-ray Radiation as Source of Short pulses of X-ray Beam
9:30	A. Zholents, LBNL	Generation of Femto-Second X-ray Pulses at Synchrotron Sources

10:05	Refreshment Break	ITP Front Patio

Convener: K. von Bibber

10:25	D. Cline, UCLA	Advanced Linear Accelerator Development for Linac Ring Colliders

Snowmass96

11:00	J. Wurtele, LBL	Summary of Accelerator Issues

Beam Sources - Status, Advances, and Limitations:

11:40	K. Leung, LBL	High Current Short Pulse Ion Sources
12:20	Lunch	ITP Patio

Convener: B. Zotter

1:45	R. Macek, LANL	High Intensity Neutron Spallation Sources

Report on New Advances:

2:25	W. Molzon, UCI	High Intensity Muon Source
3:05	C. Bula, Princeton Univ.	Observation of Nonlinear QED Effects in Electron-Laser Collisions
3:25	Refreshment Break	ITP Patio
4:00	M. Ottinger, Univ. Texas	SYN2 - A New Model for Tracking Space Charge Perturbed Synchrotron Beams
4:20	Z. Parsa, C. Pellegrini A. Sessler, R. Siemann, J. Wurtele, V. Telnov, T. Marshall, B. Gorev, R. Macek, K. Leung, D. Cline, A. Skrinsky and others. TBA	Reports* on New Advances and Round Table Discussion on Basic Issues
5:45	Beer and Chips	ITP Front Patio
6:15	Fiesta Barbeque	ITP Front Patio

Friday, August 23, 1996
Convener: N. Andreev

Time	Speaker	Title
Reports on New Advances (continued:		
9:00 am	J. Allen, UCSB	Photon Assisted Transport in Semiconductor Quantum Structures with the UCSB FEL's
9:45	J. Wells, Harvard Univ.	Superintense Laser - Atom Interactions:
10:15	Refreshment Break	ITP Front Patio
10:35	W. Lee, Princeton Univ.	Perturbative Particle Simulation of Space Charge Effects for a K-V Beam
11:05	M. Pato, Univ. Sao Paulo	
12:15 pm	Lunch	ITP Front Patio
1:45		Reports* (continues)
	Z. Parsa	Summary and Closing Talk

LIST OF PARTICIPANTS*

James Allen	University of California, Santa Barbara
Nikolai Andreev	Russian Academy of Sciences, IVTAN, HEDRC
Christopher Barty	University of California, San Diego
Boris Breizman	University of Texas, Austin
Christian Bula	Stanford Linear Accelerator Center
Pisin Chen	Stanford Linear Accelerator Center
David Cline	University of California, Los Angeles
Alex Dragt	University of Maryland
Eric Esarey	Naval Research Laboratory
Jorge Fontana	Santa Barbara, CA
Vladimir Gorev	Kurchatov Institute
Martin Gunderson	University Southern California
Samuel Heifets	Stanford Linear Accelerator Center
Forrest Jobes	Princeton University
Chan Joshi	University of California, Los Angeles
Thomas Katsouleas	University Southern California
Wayne Kimura	STI Optronics, Inc.
Wei-li Lee	Princeton University
Ka-Ngo Leung	Lawrence Berkeley Laboratory
Konstantin Lotov	Budker Institute of Nuclear Physics
Robert Macek	Los Almos National Laboratory
William Marciano	Brookhaven National Laboratory
Thomas Marshall	Columbia University
William Molzon	University of California, Irvine
Warren Mori	University of California, Los Angeles
Gerard Mourou	University of Michigan
Michael Ottinger	University of Texas, Austin
Richard Pantell	Portola Valley, CA
Zohreh Parsa	Brookhaven National Laboratory
Mauricio Pato	Univ. de São Pualo
Claudio Pellegrini	University of California, Los Angeles
Robert Phillips	Stanford Linear Accelerator Center
Donald Rule	Naval Surface Warfare Center
Andrew Sessler	Lawrence Berkeley National Laboratory
William Sharp	Lawrence Livermore National Laboratory
Alexander Shchagin	Kharkov Inst. of Physics & Tech.
Robert Siemann	Stanford Linear Accelerator Center
Alexander Skrinsky	Budker Institute of Nuclear Physics
Gennady Stupakov	Stanford Linear Accelerator Center
Valery Telnov	Budker Institute for Nuclear Physics
Donald Umstadter	University of Michigan
Karl van Bibber	Lawrence Livermore National Laboratory

*This may not include the names of the late resistants.

Jack Wells	Harvard University
Jonathan Wurtele	University of California, Berkeley
Zafar Yasin	King Fahd Univ. of Petroleum & Minerals
Simon Yu	Lawrence Berkeley National Laboratory
Alexander Zholents	Lawrence Berkeley National Laboratory
Bruno Zotter	CERN, Switzerland

Author Index

A
Andreev, N. E., 61

B
Breizman, B. N., 75
Bula, C., 165

C
Chebotaev, P. Z., 75
Chen, P., 95
Cline, D. B., 145

F
Fontana, J. R., 31

G
Gorbunov, L. M., 61

H
Hsu, J. L., 21

K
Katsouleas, T., 21
Kimura, W. D., 31
Kirsanov, V. I., 61
Kudryavtsev, A. M., 75

L
Leung, K.-N., 155
Lotov, K. V., 75

M
Marshall, T. C., 105
Mori, W. B., 21

N
Noble, R. J., 95

P
Pantell, R. H., 53
Parsa, Z., 135, 179
Pato, M. P., 179
Phillips, R. M., 11

S
Sakharov, A. S., 61
Schroeder, C. B., 21
Shchagin, A. V., 135
Siemann, R. H., 1
Skrinsky, A. N., 41, 75

T
Telnov, V., 121

U
Umstadter, D., 89

W
Wurtele, J. S., 21

Z
Zhang, T. B., 105

AIP Conference Proceedings

	Title	L.C. Number	ISBN
No. 334	Few-Body Problems in Physics (Williamsburg, VA 1994)	95-76481	1-56396-325-6
No. 335	Advanced Accelerator Concepts (Fontana, WI 1994)	95-78225	1-56396-476-7 (Set) 1-56396-474-0 (Book) 1-56396-475-9 (CD-Rom)
No. 336	Dark Matter (College Park, MD 1994)	95-76538	1-56396-438-4
No. 337	Pulsed RF Sources for Linear Colliders (Montauk, NY 1994)	95-76814	1-56396-408-2
No. 338	Intersections Between Particle and Nuclear Physics 5th Conference (St. Petersburg, FL 1994)	95-77076	1-56396-335-3
No. 339	Polarization Phenomena in Nuclear Physics Eighth International Symposium (Bloomington, IN 1994)	95-77216	1-56396-482-1
No. 340	Strangeness in Hadronic Matter (Tucson, AZ 1995)	95-77477	1-56396-489-9
No. 341	Volatiles in the Earth and Solar System (Pasadena, CA 1994)	95-77911	1-56396-409-0
No. 342	CAM-94 Physics Meeting (Cacun, Mexico 1994)	95-77851	1-56396-491-0
No. 343	High Energy Spin Physics Eleventh International Symposium (Bloomington, IN 1994)	95-78431	1-56396-374-4
No. 344	Nonlinear Dynamics in Particle Accelerators: Theory and Experiments (Arcidosso, Italy 1994)	95-78135	1-56396-446-5
No. 345	International Conference on Plasma Physics ICPP 1994 (Foz do Iguaçu, Brazil 1994)	95-78438	1-56396-496-1
No. 346	International Conference on Accelerator-Driven Transmutation Technologies and Applications (Las Vegas, NV 1994)	95-78691	1-56396-505-4
No. 347	Atomic Collisions: A Symposium in Honor of Christopher Bottcher (1945-1993) (Oak Ridge, TN 1994)	95-78689	1-56396-322-1
No. 348	Unveiling the Cosmic Infrared Background (College Park, MD, 1995)	95-83477	1-56396-508-9

Title	L.C. Number	ISBN
No. 349 Workshop on the Tau/Charm Factory (Argonne, IL, 1995)	95-81467	1-56396-523-2
No. 350 International Symposium on Vector Boson Self-Interactions (Los Angeles, CA 1995)	95-79865	1-56396-520-8
No. 351 The Physics of Beams Andrew Sessler Symposium (Los Angeles, CA 1993)	95-80479	1-56396-376-0
No. 352 Physics Potential and Development of $\mu^+\mu^-$ Colliders: Second Workshop (Sausalito, CA 1994)	95-81413	1-56396-506-2
No. 353 13th NREL Photovoltaic Program Review (Lakewood, CO 1995)	95-80662	1-56396-510-0
No. 354 Organic Coatings (Paris, France, 1995)	96-83019	1-56396-535-6
No. 355 Eleventh Topical Conference on Radio Frequency Power in Plasmas (Palm Springs, CA 1995)	95-80867	1-56396-536-4
No. 356 The Future of Accelerator Physics (Austin, TX 1994)	96-83292	1-56396-541-0
No. 357 10th Topical Workshop on Proton-Antiproton Collider Physics (Batavia, IL 1995)	95-83078	1-56396-543-7
No. 358 The Second NREL Conference on Thermophotovoltaic Generation of Electricity	95-83335	1-56396-509-7
No. 359 Workshops and Particles and Fields and Phenomenology of Fundamental Interactions (Puebla, Mexico 1995)	96-85996	1-56396-548-8
No. 360 The Physics of Electronic and Atomic Collisions XIX International Conference (Whistler, Canada, 1995)	95-83671	1-56396-440-6
No. 361 Space Technology and Applications International Forum (Albuquerque, NM 1996)	95-83440	1-56396-568-2
No. 362 Two-Center Effects in Ion-Atom Collisions (Lincoln, NE 1994)	96-83379	1-56396-342-6
No. 363 Phenomena in Ionized Gases XXII ICPIG (Hoboken, NJ, 1995)	96-83294	1-56396-550-X
No. 364 Fast Elementary Processes in Chemical and Biological Systems (Villeneuve d'Ascq, France, 1995)	96-83624	1-56396-564-X

	Title	L.C. Number	ISBN
No. 365	Latin-American School of Physics XXX ELAF Group Theory and Its Applications (México City, México, 1995)	96-83489	1-56396-567-4
No. 366	High Velocity Neutron Stars and Gamma-Ray Bursts (La Jolla, CA 1995)	96-84067	1-56396-593-3
No. 367	Micro Bunches Workshop (Upton, NY, 1995)	96-83482	1-56396-555-0
No. 368	Acoustic Particle Velocity Sensors: Design, Performance and Applications (Mystic, CT, 1995)	96-83548	1-56396-549-6
No. 369	Laser Interaction and Related Plasma Phenomena (Osaka, Japan 1995)	96-85009	1-56396-445-7
No. 370	Shock Compression of Condensed Matter-1995 (Seattle, WA 1995)	96-84595	1-56396-566-6
No. 371	Sixth Quantum 1/f Noise and Other Low Frequency Fluctuations in Electronic Devices Symposium (St. Louis, MO, 1994)	96-84200	1-56396-410-4
No. 372	Beam Dynamics and Technology Issues for + - Colliders 9th Advanced ICFA Beam Dynamics Workshop (Montauk, NY, 1995)	96-84189	1-56396-554-2
No. 373	Stress-Induced Phenomena in Metallization (Palo Alto, CA 1995)	96-84949	1-56396-439-2
No. 374	High Energy Solar Physics (Greenbelt, MD 1995)	96-84513	1-56396-542-9
No. 375	Chaotic, Fractal, and Nonlinear Signal Processing (Mystic, CT 1995)	96-85356	1-56396-443-0
No. 376	Chaos and the Changing Nature of Science and Medicine: An Introduction (Mobile, AL 1995)	96-85220	1-56396-442-2
No. 377	Space Charge Dominated Beams and Applications of High Brightness Beams (Bloomington, IN 1995)	96-85165	1-56396-625-7
No. 378	Surfaces, Vacuum, and Their Applications (Cancun, Mexico 1994)	96-85594	1-56396-418-X
No. 379	Physical Origin of Homochirality in Life (Santa Monica, CA 1995)	96-86631	1-56396-507-0
No. 380	Production and Neutralization of Negative Ions and Beams / Production and Application of Light Negative Ions (Upton, NY 1995)	96-86435	1-56396-565-8
No. 381	Atomic Processes in Plasmas (San Francisco, CA 1996)	96-86304	1-56396-552-6

Title	L.C. Number	ISBN
No. 382 Solar Wind Eight (Dana Point, CA 1995)	96-86447	1-56396-551-8
No. 383 Workshop on the Earth's Trapped Particle Environment (Taos, NM 1994)	96-86619	1-56396-540-2
No. 384 Gamma-Ray Bursts (Huntsville, AL 1995)	96-79458	1-56396-685-9
No. 385 Robotic Exploration Close to the Sun: Scientific Basis (Marlboro, MA 1996)	96-79560	1-56396-618-2
No. 386 Spectral Line Shapes, Volume 9 13th ICSLS (Firenze, Italy 1996)		1-56396-656-5
No. 387 Space Technology and Applications International Forum (Albuquerque, NM 1997)	96-80254	1-56396-679-4 (Case set) 1-56396-691-3 (Paper set)
No. 388 Resonance Ionization Spectroscopy 1996 Eighth International Symposium (State College, PA 1996)	96-80324	1-56396-611-5
No. 389 X-Ray and Inner-Shell Processes 17th International Conference (Hamburg, Germany 1996)	96-80388	1-56396-563-1
No. 390 Beam Instrumentation Proceedings of the Seventh Workshop (Argonne, IL 1996)	97-70568	1-56396-612-3
No. 391 Computational Accelerator Physics (Williamsburg, VA 1996)	97-70181	1-56396-671-9
No. 392 Applications of Accelerators in Research and Industry: Proceedings of the Fourteenth International Conference (Denton, TX 1996)	97-71846	1-56396-652-2
No. 393 Star Formation Near and Far Seventh Astrophysics Conference (College Park, MD 1996)	97-71978	1-56396-678-6
No. 394 NREL/SNL Photovoltaics Program Review Proceedings of the 14th Conference— A Joint Meeting (Lakewood, CO 1996)	97-72645	1-56396-687-5
No. 395 Nonlinear and Collective Phenomena in Beam Physics (Arcidosso, Italy 1996)	97-72970	1-56396-668-9
No. 396 New Modes of Particle Acceleration— Techniques and Sources (Santa Barbara, CA 1996)	97-72977	1-56396-728-6

PHYSICS LIBRARY
Imperial College, London SW7 2BZ

This book may be recalled after 2 weeks if required by another reader